INTRODUCING THE BRAIN

CONTENTS

2 Introducing the brain

4 Brain parts

6 What's inside

8 Baby brains

10 Intelligence

12 Eyesight

14 Memory

16 Remembering

18 Staying the same

20 Consciousness

22 Mental illness

24 Drugs

25 Out of this world?

26 Animal brains

28 Computer brains

30 Brains in history

32 A glossary

This part of the book is all about the brain. Your brain is just over 1kg (nearly 3lbs) of gooey, slimy, wobbly gelatinous stuff which smells of blue cheese. It sounds revolting, but it is the most vital organ in your body.

The brain rules supreme. Perched above your neck, beneath your skull, it controls almost all your activities: thinking, feeling, talking, moving, and just keeping alive. Without it, you wouldn't really be human at all.

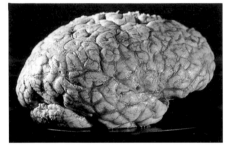

A human brain

Your brain operates 24 hours a day. It never gets tired. It is your own living life-support machine.

Without your brain you wouldn't be able to do any of the things shown in these pictures.

Taste

Smell

Hear

For a link to a website where you can watch a short video about the brain, go to **www.usborne-quicklinks.com**

OLOGISTS

Different sorts of scientists study the brain in different ways.

Neurologists study the cells in the brain and the nervous system (see page 7).

Psychologists study how humans behave.

Craniologists study the shape and size of the human skull.

Psychiatrists study what happens when the brain goes wrong and people act in strange ways.

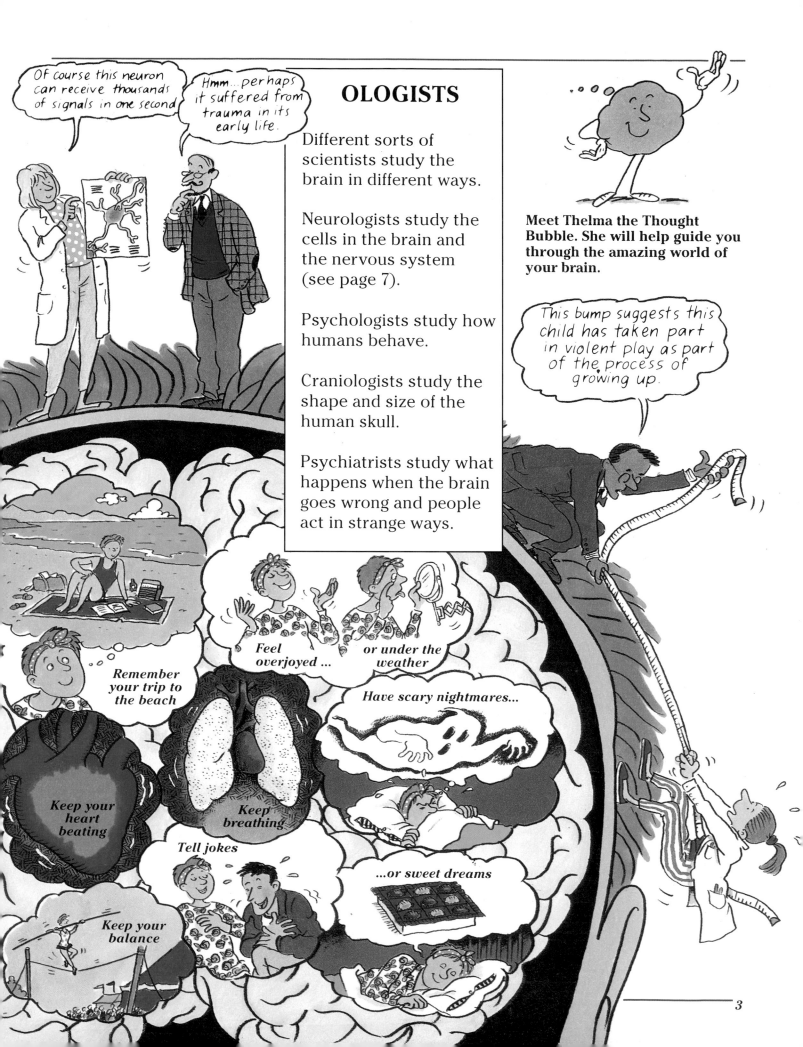

Of course this neuron can receive thousands of signals in one second

Hmm...perhaps it suffered from trauma in its early life.

Meet Thelma the Thought Bubble. She will help guide you through the amazing world of your brain.

This bump suggests this child has taken part in violent play as part of the process of growing up.

Remember your trip to the beach

Feel overjoyed ...

or under the weather

Have scary nightmares...

Keep your heart beating

Keep breathing

Tell jokes

...or sweet dreams

Keep your balance

BRAIN PARTS

Different parts of your brain control all the different things that happen in your body. The picture below shows you which part does what. Each part is a different shade, but in real life your brain parts (when the blood's washed off them) are pinkish grey or dirty white. The top of your brain is divided into two domes, called cerebral hemispheres. This makes it look like a large wrinkly walnut.

BRAINMAP
The cortex is the part of the brain you use to think and feel. It is the part that makes you aware of what you are doing. This "map" of the left side of the cortex shows what some of the areas control.

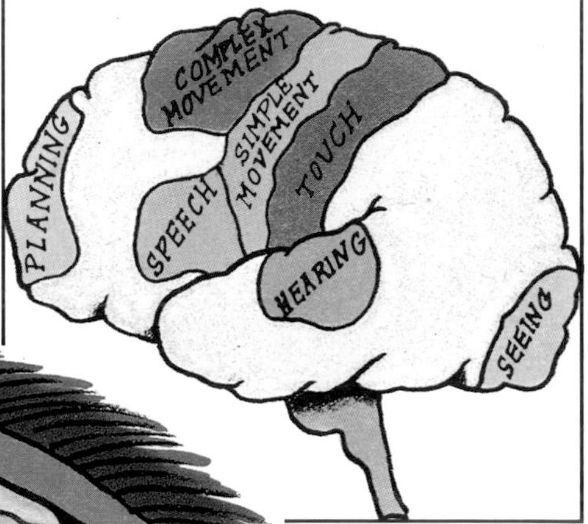

PLANNING
COMPLEX MOVEMENT
SPEECH
SIMPLE MOVEMENT
TOUCH
HEARING
SEEING

The two cerebral hemispheres together make up the **cerebrum.** *The outer layer of the cerebrum is called the* **cortex.**

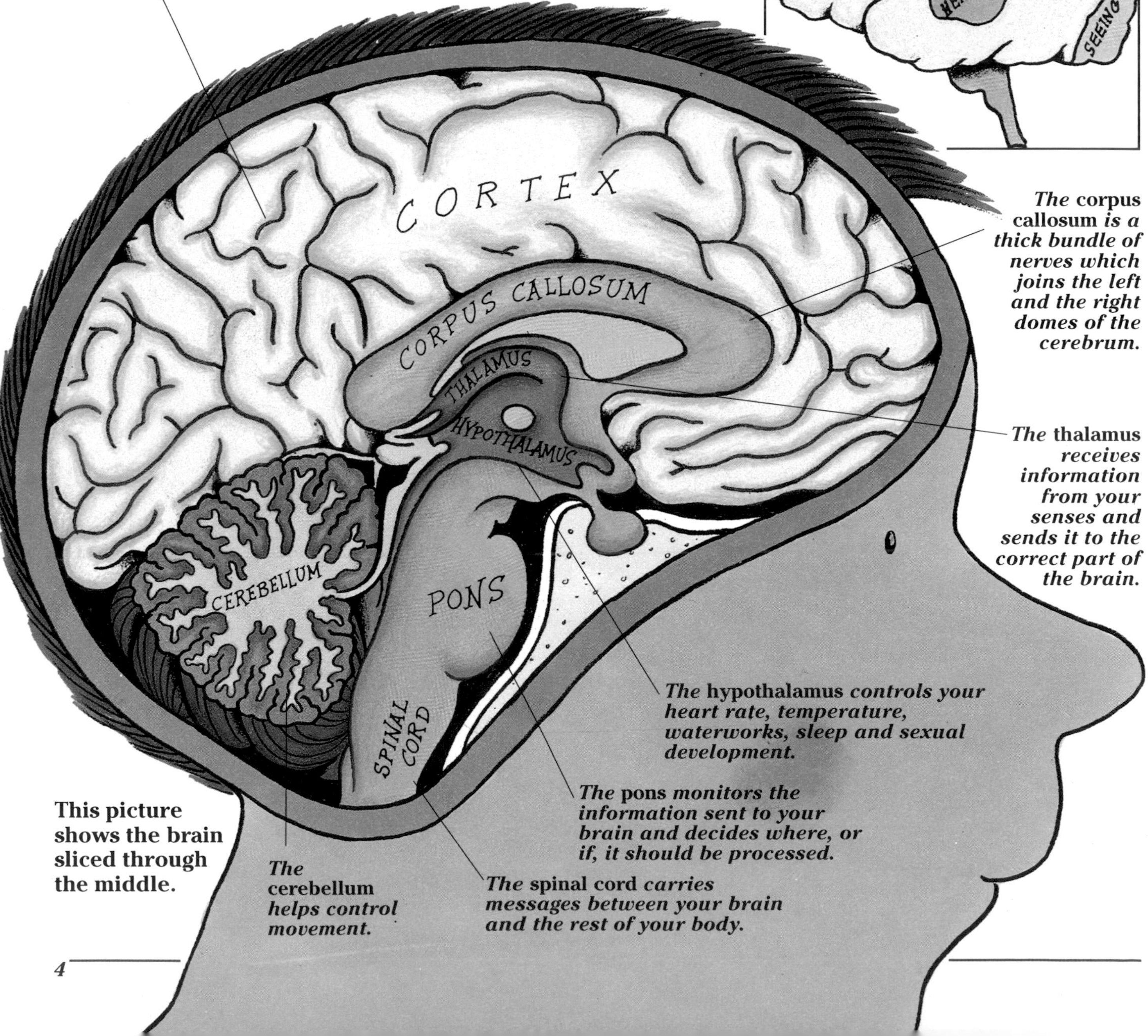

CORTEX

CORPUS CALLOSUM

THALAMUS

HYPOTHALAMUS

CEREBELLUM

PONS

SPINAL CORD

The **corpus callosum** *is a thick bundle of nerves which joins the left and the right domes of the cerebrum.*

The **thalamus** *receives information from your senses and sends it to the correct part of the brain.*

The **hypothalamus** *controls your heart rate, temperature, waterworks, sleep and sexual development.*

The **pons** *monitors the information sent to your brain and decides where, or if, it should be processed.*

This picture shows the brain sliced through the middle.

The **cerebellum** *helps control movement.*

The **spinal cord** *carries messages between your brain and the rest of your body.*

HEMISPHERES

Each side of your brain, or cerebral hemisphere, looks after the opposite side of the body. Each side is also in charge of different kinds of thoughts and actions.

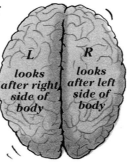

L looks after right side of body

R looks after left side of body

The left side is used for speech and language. It is also used for other tasks which require you to do things in a particular order, such as sums or tying a shoelace.

The right side is used for thinking in pictures. If you had to draw a map of the way to school, you would picture the route in your head with the right hemisphere.

Your corpus callosum lets one hemisphere know what the other is doing. Without it, you could read and understand the word "pig" (using your left hemisphere) but would not be able to picture a pig in your mind (which uses the right hemisphere).

RIGHT OR LEFT?

Answer each question and try to figure out which side of your brain it is testing. Solutions on page 98.

1. Which of the boxes a), b), c) or d) can be made by folding the flat piece of paper on the right?

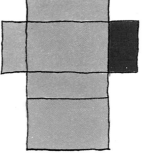

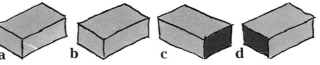

a   b   c   d

2. Which number comes next in the series: 17, 14, 11, 8...(7, 3, 5 or 0)

3. Which is the odd one out?

a   b   c   d   e

4. If Dot goes with Jemma, who does Rosie go with?

Dot   Jemma   Rosie

Anna   Kathy

Becky   or   Mary?

PONS TO THE RESCUE

Have you ever walked into a room where there is a really smelly piece of cheese?

At first, the smell can be completely overwhelming and almost unbearable.

But, after being in the room only a few minutes, you begin to stop noticing it.

The smell hasn't gone, but the pons has stopped sending on the smelly messages to be processed.

WHAT'S INSIDE?

No one understands exactly how the brain works. But scientists know the answer lies with the billions of tiny cells, called neurons or nerve cells, which make up your brain. All your feelings, thoughts and actions are caused by electrical and chemical signals passing from one neuron to the next. It may seem incredible, but a complicated feeling such as jealousy is a series of electrical and chemical changes.

WHAT DOES A NEURON LOOK LIKE?

A neuron looks a little like a tiny octopus, but with many more tentacles (some have several thousand). In all the different parts of your brain, neurons are carrying signals which allow you to move, hear, see, taste, smell, remember, feel and think.

Simple neurons magnified 1000 times

The cell body. Controls the cell and directs all its activites.

Signal passing between neurons.

Dendrites. Tentacles which radiate from the cell body. They receive signals from axons and carry them to the cell body.

Axon. A larger tentacle, often with branching ends, which carries signals away from the cell body and passes them on to dendrites of other neurons.

Some axons are long enough to stretch across the brain or even all the way down the spinal cord.

HOW DO NEURONS CARRY MESSAGES?

At football games, people sometimes do "The Wave", throwing their arms in the air one after the other. A wave of arms travels from one end of a row to the other. Messages travel down neurons in a similar way. But instead of arms being thrown in the air, tiny pulses of electricity are fired off, one after another, down the length of the axon.

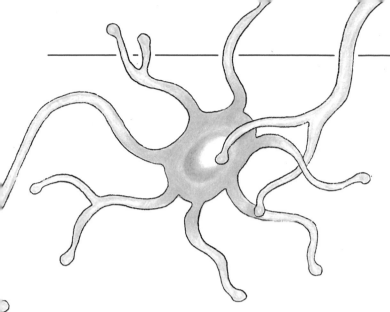

JUMPING THE GAP

Axons and dendrites are separated by tiny gaps called synapses. When a signal reaches the end of an axon, special chemicals are released which spread across the gap. When the chemicals reach the other side, the dendrite fires off an electrical pulse.

STARTLING STATISTICS

• **The fastest brain messages can travel at about 580 kmph (360 mph).**

• **You have about 100 billion neurons. Each one can be connected to thousands of others. This means there are trillions of different routes a message can take around your brain.**

• **Each brain cell may receive hundreds and thousands of incoming signals every second.**

FEEDING THE BRAIN

Your body needs oxygen like a car needs fuel. Oxygen is carried around the body in the blood. Different parts of your body use up different amounts of oxygen, depending on how much they do. The brain is so active it uses almost a quarter of the body's oxygen, although it is only 2% of its total weight.

GREY MATTER

Some people say "Use your grey matter" if they want you to think very hard. Grey matter is what makes up most of your cortex. It consists of millions of cell bodies packed tightly together. Much of the rest of the cerebrum is made up of bundles of axons. This is called white matter.

THE NERVOUS SYSTEM

The nervous system is a network of neurons stretching from your brain to the tips of your toes. Some neurons send messages to the brain about what's happening inside and outside the body. The brain decides what should be done. Instructions are then sent back down other neurons, via the spinal cord, to muscles, organs or cells, which carry out the response.

If the brain receives information about chocolate cake, it will send a message to your arm to grab it.

Pathways of neurons

For a link to a website where you can probe a virtual brain and move different parts of the body, go to **www.usborne-quicklinks.com**

BABY BRAINS

Babies are born with a few, very limited, responses: they can turn their cheek if it is touched, hear and smell, find things to suck, and see black and white patterns. Yet, within only a few days, they can do something as complex as recognizing their mother's face. Their brains have already begun to analyze the outside world. Babies and young children absorb enormous amounts of information every day. You probably learn more in your first five years than in the whole of the rest of your life.

TRIAL AND ERROR

Babies learn by a process of trial and error. As they explore their surroundings, they gradually understand more and more about how the world works.

The picture strip above shows how a baby discovers the link between shaking a rattle and the nice jangly sound it makes.

It takes the baby several chance shakings before he discovers the connection between the rattle and the sound.

These babies are learning as they investigate their world.

BABY TALK

One of the most scientifically baffling things that babies learn to do is to talk.

Between the ages of one and two, a baby learns to utter a few words and to understand simple sentences.

Of course... SUCK...SUCK... NIETZSCHE advanced a doctrine... SUCK...SUCK... of Eternal Recurrence...

Between the ages of two and five, a child learns about ten words a day. (Anyone who has tried to learn a foreign language will know this is an enormous amount.) In three years, a child's vocabulary increases from a few hundred words to as many as 15,000.

OUT OF SIGHT...

The test below was carried out on a nine-month-old baby. It made pyschologists believe that if a young baby couldn't see an object she thought it no longer existed.

The baby is shown a car. She tries to grab it. In full view of the baby, the car is hidden under a cloth. The baby loses interest and doesn't try to get the car.

A later test proved this theory wrong. It showed that a baby does know a thing exists even when she can't see it - but she doesn't think she has any control over it.

The baby tracks a toy elephant moved in front of her face. When the elephant disappears behind a screen, the baby tracks its movements until it reappears. If the elephant is exchanged for a giraffe behind the screen, the baby looks shocked, often bursting into tears. This shows she knew the elephant was still there, even though she could not see it.

HOW THE BRAIN GROWS

Neurologists think that you don't produce any new brain cells after you are born. So brains don't get bigger by adding more neurons. Instead, your brain grows by increasing the number of the connections between neurons. Axons grow new branches which link up with dendrites. As the number of connections between neurons increases, you become capable of more and more complex kinds of thinking.

Any brain cells you lose through a bang on the head can't be replaced. But, with over 100 billion, you can afford to lose a few thousand.

LEARNING TO THINK

As young children, our thinking power is limited. It takes time to understand how the world works.

This child of four agrees that these two beakers hold the same amount of water.

If the water in one beaker is poured into a thinner beaker, in front of the child, she now says the thinner beaker holds more water. A child of seven wouldn't make this mistake.

b a   a b

If this piece of string were straightened out, where would the ends reach: points *a* or points *b*?

Most children under five think the ends would stay in the same place. This is because they cannot yet picture changes in their heads. Most older children know the answer is *b*.

INTELLIGENCE

What makes one person a genius and another just plain average? It is probably a mixture of the brain they are born with and the training they receive. Different people are intelligent in different ways: one may be good at French, but hopeless at chess, while another may understand feelings, but not understand numbers.

BIG HEADS

Some scientists have claimed that big skulls mean bigger brains, and that bigger brains mean brighter people. Men, women and people of different races do have different sized brains. But there is no proof that this has anything to do with how intelligent they are.

TESTING TIMES

In 1905, a Frenchman called Alfred Binet made up some tests to measure intelligence. His tests were meant only to include questions which did not need any special learning. Similar tests, called IQ tests, are still used today. Some people think IQ tests are unfair, because children who are used to doing tests like this usually get higher scores.

BRAIN BOX

Ruth Lawrence was a child genius. At seven, she was taking tests meant for 18-year-olds, and at 11 she was studying at Oxford University. Most of the other students were twice her age.

TAKING THE TEST

IQ tests are made up of different sorts of puzzles and questions. Some use words and numbers, while others use patterns and shapes. This is so that different kinds of intelligence, and both your right and left cerebral hemispheres (see page 4), are tested. Try these tests and see how you do. You can find the answers on page 98.

1. How many triangles are there in this picture? Some smaller triangles may make up bigger triangles.

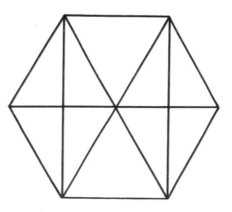

2. Fill in the missing number.

S	H	I	P
19/4	8/15	9/?	16/11
D	O	C	K

3. Which of these shapes will complete the square?

BORN OR BRED?

Inside all your cells are tiny chains of chemicals, called genes. Genes carry instructions which control how your body works. If you were born with a fixed level of intelligence, it would be controlled by your genes.

No two people have exactly the same genes - except for identical twins. If intelligence comes from genes, identical twins should be equally intelligent. This means they should have similar IQs, even if brought up apart.

When psychologists looked at separated twins' IQ tests, they found that they had a high chance of having much the same result. This backs up the argument that at least part of intelligence comes from your genes.

Nick and Paddy are identical twins. They live apart and grew up doing completely different things. See what happened when they both took an IQ test.

4.

Fill in the missing number.

5. Which of the following words means the <u>same</u> as or the <u>opposite</u> of Tall? (Handsome, Dark, Thin, Short, Fat)

6. Beetle/Insect; Sparrow/.... ? (Ant, Dove, Slug, Bird, Feathers).

7.

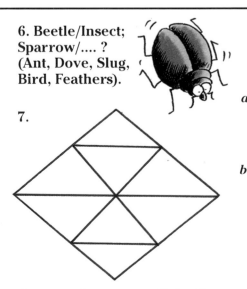

Can you draw over all the lines in this picture without going over any part of a line twice?

8. Which piece from row *b* makes row *a* a complete set?

a

b

9. In a box, there are 3 pairs of red boots and 2 pairs of blue boots. How many boots would you have to take out, without seeing them, to be certain of getting a complete pair?

EYESIGHT

Your sense organs receive information from the outside world and turn it into electrical signals. These signals are sent to your brain, where they are interpreted as sights, sounds, smells, tastes and feeling. These two pages show how your eyes and brain work together to let you see.

SEEING

When you see an object, light from it travels into your eye and a 2-D flat image is projected onto your retina (which is like a curved screen at the back of your eye). This image is converted into a series of electrical signals by cells called rods and cones. The signals travel to your brain, where they are interpreted as 3-D images. You can read more about eyes and seeing on pages 38 to 41.

Real 3-D world

Cross-section through an eye.

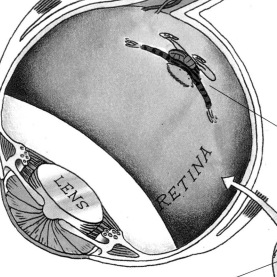

LENS

RETINA

Brain processes and analyzes electrical pulses and creates a 3-D image.

Optic nerve - bundle of nerves down which electrical pulses travel to the brain.

Flat image of world (known as the retinal image) projected onto retina.

Rods and cones - cells which convert the image on your retina into pulses of electricity.

A 3-D WORLD

The image on your retina is 2-D, but you see in 3-D. This is partly because you have two eyes, each of which gives you a slightly different view of an object. This is called binocular vision. Read about it on page 39. Your brain also analyzes the retinal image and uses key elements within it to build up a 3-D world in your head.

This picture has the same key elements as the flat image on your retina.

○ *Size. Similar objects of different sizes are interpreted as being at different distances.*

○ *Arrows. Arrow-shaped lines are interpreted as inside or outside corners.*

○ *Overlapping. If one object obscures another, you see the overlapping object as being nearer.*

○ *Lines. Your brain realizes that parallel lines appear to move closer together as they get farther away.*

FUNNY PHOTOS

Because your brain interprets retinal images so quickly, you are not usually aware of their actual size.

In this photo (which is flat like a retinal image) you see two girls. They look the same size, but at different distances from you.

In this photo, the image of the girl who is farther away has been pasted next to the girl in the foreground. She looks much smaller than she did in the photo above.

OPTICAL ILLUSIONS

Optical illusions are the brain making a wrong guess. They tell us how your brain normally analyzes images.

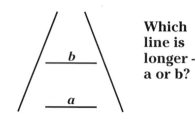

Which line is longer - a or b?

Line *b* looks longer, but *a* and *b* are the same size. Your brain treats the converging lines as parallel, so thinks line *b* is farther away than line *a*. So as lines *a* and *b* project the same-sized retinal images, your brain guesses that line *b* is longer.

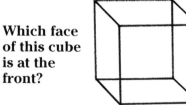

Which face of this cube is at the front?

As you look, the cube seems to flip. You don't have enough elements to figure out which way around it is. So your brain makes two guesses, but cannot choose between them.

MORE THAN MEETS THE EYE

Seeing is a lot more than just looking. What you see depends on what you know, expect and want, as well on the image on your retina.

I
WANT TO BE
BE A SUPERSTAR

When you read the words above, you probably didn't notice the double 'be' because your brain didn't expect it to be there.

The pictures in the middle of this series are distorted. Depending on which way you look at the series, you will either see a distorted face or a distorted woman.

WHITER THAN WHITE

Hold the book up in front of your face and stare at this set of broken rays. As you look, a white ring will seem to appear. It will look whiter than the rest of the white page.

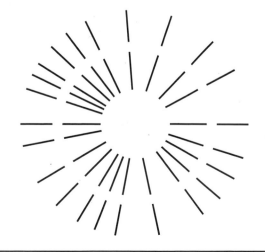

Of course this isn't the case. This illusion occurs because your brain assumes that there must be a solid white circle in front of the rays, which is covering up part of them.

MEMORY

LONG-TERM MEMORY

Life without memory would be impossible. You don't just need your memory to recall a telephone number or the date of your best friend's birthday. You need it to remember who you are, how to walk, how to speak and whether or not you like Irish stew. In fact it is your memory that makes you who and what you are.

Everybody has two types of memory: long-term and short-term. Nothing stays in your short-term memory for more than a few minutes. Anything that you can remember for more than that is in your long-term memory. Things can stay in your long-term memory for hours, weeks, months, years or even for the rest of your life.

Your long-term memory stores everything you know. By the time you are eight years old, it holds more information than a million encyclopedias.

Some of the things in your long-term memory.

DRAGON SKATEBOARD GEOGRAPHY

A vocabulary of more than 15,000 words.

How to tie your shoelaces...

and ride a bicycle without falling off.

The times of, and characters in, the TV shows you like best.

The names of all your class at school.

What you had for dinner last night.

How to find your way home from school.

Who won the World Cup.

How to read, write and add.

This is just the tiniest fraction of what is in your long-term memory. Amazingly, it will never become full, because it has a limitless capacity. This means it could go on storing new information, even if you were to live until you were over a hundred.

For a link to a website where you can test your memory with online activities, go to www.usborne-quicklinks.com

SHORT-TERM MEMORY

On the other hand, your short-term memory can only store a maximum of nine things at one time. Most people cannot manage more than seven.

You can test this yourself. Repeat the list of numbers below to several friends. Ask them to write down, in the right order, as many of them as they can remember.

Most people can only remember between five and seven numbers.

After a few minutes, facts in your short-term memory are displaced by new ones. The old facts either just fade away or are transferred to your long-term memory.

STORAGE

You store different sorts of information in your memory in different ways. Most facts in your short-term memory are stored as sounds.

Read the first sequence of letters below.

R-J-L-T-M-X-S-Q-F

Write down as many as you can remember in the right order.

Now do the same with the second set of letters.

B-C-T-G-E-P-D-V-G

Most people do better with the first set than the second. This is because in the second set the letters sound similar (bee-cee-tee-etc.), so they get confused more easily.

In your long-term memory, words are usually stored according to what they mean, not what they sound like. If a teacher says "School is closed next week," you would not remember if she had said that particular sentence or another with the same meaning: "Next week, school is shut."

You can also store sights, sounds and smells in your long-term memory. This means you can recognize a famous painting, hum familiar tunes to yourself, and know when your sister's borrowed your mother's perfume.

WHAT ARE MEMORIES?

Everything you learn and all your experiences are encoded in your brain as patterns of electrical pulses passing between neurons.

Memories are patterns of pulses which are repeated without the experience actually taking place. A particular memory returns each time a pattern of electrical pulses is activated.

REMEMBERING

There are some things you never forget, such as your name and age. Other facts and events - such as your worst day at school or your best birthday ever - you can remember whenever you want. But other things can be harder to recall. Remembering can be made easier if you use hints and cues. You can also help yourself by the way you first learn information.

It's easier to remember a fact or event if you are in the place where you first stored it.

Just thinking about the place where you learned something can help too.

Large amounts of information can be very difficult to recall. The way you first learn it can make remembering easier.

Organize the information into groups, and give each one a heading. This is like making a filing cabinet in your head.

When you have to recall the facts, just remembering the headings will make the information easier to recall.

Your memory is full of all sorts of information that, most of the time, you don't even know you have stored.

You would only become aware of it if you were given a strong enough reminder or cue which brings it back to you.

Remembering can be painful. There may be some things that you had hoped to block out of your memory altogether.

For a link to a website where you can play a game to see if you would be a reliable witness, go to www.usborne-quicklinks.com

LOONY LISTS

Imagine you are going shopping tomorrow, and a friend gives you a list of things she wants you to buy for her. You know you are always losing lists, so you want to commit it to memory. If you try learning it by heart, the chances are you will have forgotten a few things by the next day. But if you try to give the list meaning (the sillier the better), you will find it much easier.

One way of doing this is to make all the items on the list part of a story. Another way is to imagine you are walking through your house. As you enter each room you "place" several items in it in strange places or doing funny things.

Items from the shopping list above have been placed in the rooms of this house in rather odd circumstances.

Test this out with a friend. One of you tries to learn the list by heart, by repeating it to yourself. The other uses the placing method. (You could use the house here or your own.) Test yourselves after 24 hours and see who can remember the most.

SHORT-TERM TEST

You can get more in your short-term memory if you can group the facts you have to store into bigger units.

Read over the first set of letters below. Look away and recall as many as you can.

P-S-U-N-E-G-U-F-O-V-I-P-L-A

Now do the same test with the second set.

PS-UN-EG-UFO-VIP-LA

There are 14 items to store in the first set, but in the second, only six.

OVEN MITTS
GLUE
CHOCOLATE MOUSSE
COFFEEPOT
WRAPPING PAPER
BICYCLE
WOOL
BANANAS
SHOES
CAT FOOD
FEATHER DUSTERS

STAYING THE SAME

Whatever you are doing and wherever you are - whether you're sunbathing in Tahiti or skiing in Scandinavia - your brain tries to keep the conditions inside your body the same. The ability to keep the body, and the chemicals inside it, in a stable state is called homeostasis. It is controlled by the tiny part of your brain called the hypothalamus (see page 4).

HORMONES

The hypothalamus triggers the release of hormones into your blood stream. Hormones are chemicals which give instructions to your cells. Some are necessary for homeostasis, others control your growth and sexual development. Find out more about hormones on page 89.

CENTRAL HEATING

Houses with central heating usually work with a thermostat. A thermostat senses how hot or cold it is and automatically turns the radiators on or off, so that the temperature always remains constant.

The hypothalamus is your brain's thermostat. It detects changes in your body temperature, and instructs different parts of your body to heat you up or cool you down as needed.

So even though you feel hotter on a sweltering summer's day than on a icy winter's morning, if you took your temperature it would be the same on both occasions.

If you are getting too hot:

You sweat more because sweating cools down the body.

Your blood flows nearer the surface of the skin so heat can be lost.

Your muscles relax because any movement produces heat.

PHEW!

Hairs lie flat so warm air cannot be trapped next to the skin.

If you are getting too cold:

Your blood flows away from the skin so heat won't be lost through the surface.

You stop sweating.

CHATTER CHATTER CHATTER

You shiver - your muscles' quick, jerky movements produce heat.

Hairs on your body stand up so they can trap warm air next to the skin.

MONITORING THE BLOOD

Your hypothalamus is on 24-hour, around-the-clock blood alert. It is constantly monitoring your blood to make sure it has everything it needs. Here are just some of the things it controls:

OXYGEN UPTAKE

All parts of your body need oxygen to work. Oxygen is carried all over your body by your blood. If you are doing lots of work, such as running up a steep hill, you will need more oxygen than usual. Your hypothalamus will send a message to your lungs telling you to breathe more quickly, so more oxygen from the air can be taken into your lungs and, from there, into your blood.

HUNGER PANGS

You get energy to do things from your food. Food is broken down into sugar and carried in your blood to the busy parts of your body.

If you start to run out of sugar, your hypothalamus makes you feel hungry, so you eat. It also triggers the release of hormones which control how much sugar is taken up by your blood cells. As the sugar level in your blood rises, your hunger pangs disappear.

WATERWORKS AND WASTE

It's very important that your blood contains the right amount of water. With too little, your blood cells would shrivel up, but with too much they would burst.

On its journey around the body, your blood travels through your kidneys. The kidneys are like a filter. Under orders from a hormone triggered by the hypothalamus, they remove from the blood any excess water, along with any nasty waste that has collected. This water and waste make up your urine. If your blood cells need more water, your hypothalamus makes you feel thirsty, so you drink.

Cross-section through a kidney

Area of kidney where, under instructions from brain, blood is filtered.

Renal artery - carries blood into the kidney.

Renal vein - carries filtered blood out of the kidney.

Ureter - carries urine to your bladder.

19

CONSCIOUSNESS

Consciousness is what you are aware of at any one moment. It is a constantly changing state. At the moment, you are aware of what you are reading and where you are or maybe of a daydream. But you could switch to something else (what you ate for breakfast, what you plan to do tomorrow) whenever you wanted.

SCREENING

Your brain operates a screening process. Information about what's happening around you is continually entering your brain. But, unless it is considered important (like the sound of your name), it never reaches your conscious mind. This stops your consciousness from getting confused.

Lots of information entering your brain

Information that never reaches your conscious mind

Useful data in your consciousness

RUBBISH

Some of the things that may be hidden in your unconscious

UNCONSCIOUS

Sigmund Freud (1856-1939)

The psychologist Freud believed we also have an unconscious, where we hide embarrassing or painful thoughts. Things in our unconscious come out when we are unaware of what we are doing, for example, in dreams or slips of the tongue.

A FREUDIAN SLIP

Bob borrowed some money from Ted. Ted was livid, but Bob was his friend, so he tried not to think about it. After a few weeks, Ted went to see Bob. "Bob," he said "I've come about your death... I mean your debt". Freud would say Ted unconsciously wanted Bob to die.

SLEEPY HEADS AND DREAMERS

You spend more time sleeping than doing anything else. You are asleep for about one-third of your life - that could be as much as 30 years.

There are two sorts of sleep which are as different from each other as waking is from sleeping. They are called REM (Rapid Eye Movement) sleep and NREM (Non REM) sleep. During REM sleep, your brain is very active.

Activity in the brain is measured by pads, fixed to a person's head, which can pick up electrical pulses in the brain. This is called an EEG (electroencephalograph).

When you are in NREM sleep, you are deeply asleep and hard to wake up. In this state, very little happens in your brain. Throughout the night you switch between NREM and REM sleep. Most people normally fall into NREM sleep at

the beginning of the night but, after a couple of hours, they will drift into REM sleep. REM sleep is the time when you dream. The brain becomes as active as when you are awake and your eyes move quickly under your eyelids. This is where the name REM comes from.

EEG readings are recorded as a line called a trace. The closer together on the trace the peaks and troughs are, the more the activity there is taking place.

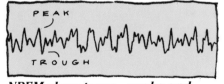

NREM sleep trace - peaks and troughs are far apart

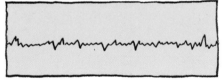

REM sleep trace - peaks and troughs are closer together.

Children dream for about 50% of the night while adults only spend about 20% of the night dreaming.

What do dreams mean and why do we have them? Freud believed that our dreams are about the things in our unconscious. But even in our sleep they cannot be expressed openly, so they are represented as symbols. So, for example, dreaming about setting out on a journey could really be a dream about dying.

Some modern psychologists think REM sleep is the time when the information we have taken in during the day is sorted out. Our memory stores are opened - new information is added to old and new categories are created. As this happens, scraps of memory, old and new, filter briefly into our conscious minds to become dreams.

This boy is dreaming about going hiking. Freud might think he was really dreaming about dying.

MENTAL ILLNESS

Your brain can go wrong, just like any other part of your body. Sometimes this can make you behave abnormally. This is called mental illness. But not all people who behave abnormally are mentally ill. Someone who is very clever is abnormal, but obviously not ill. Mental illness is always distressing and harmful to the sufferer, and to those around them. Two of the most serious forms of mental illness are schizophrenia and depression.

SCHIZOPHRENIA

Delusions of grandeur

Paranoid delusions

Hallucinations

All schizophrenia sufferers lose control of their own thinking. Some suffer from delusions. This means they believe things that are untrue. A few have delusions of grandeur. This means they think they are powerful and important, or a famous person, such as Jesus. Others have paranoid delusions and believe that people are trying to kill them or that everyone hates them.

Schizophrenics may also have hallucinations. During a hallucination, a person experiences something which isn't really there. Many hear voices telling them to do things (often dangerous) or commenting on their actions.

DEPRESSION

People with depression suffer from deep despair, hopelessness and often complete loss of energy. In some cases this alternates with periods of mania - when they appear to have boundless energy. Although on the surface people in this manic state may appear elated and happy, they are usually not in control and are often frightened and confused.

PHOBIAS

Someone who is afraid of something that most people don't find at all scary, is said to have a phobia. Some phobias can seriously interfere with everyday life.

These are some more unusual phobias:

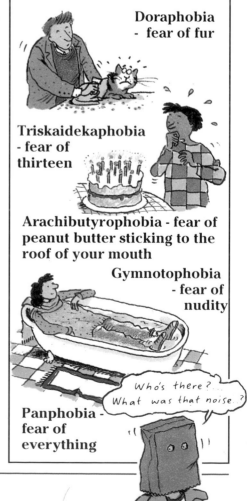

Doraphobia - fear of fur

Triskaidekaphobia - fear of thirteen

Arachibutyrophobia - fear of peanut butter sticking to the roof of your mouth

Gymnotophobia - fear of nudity

Panphobia - fear of everything

For a link to a website where you can find out more about phobias, go to www.usborne-quicklinks.com

CAUSES AND CURES

Mental illness is probably caused by a combination of abnormalities in the genes you are born with and things that happen in your life (environmental factors). Doctors have two main ways of treating mental illness - biological and psychological treatments.

Biological treatments look at what happens to the chemicals in the brain during mental illness and try to return them to normal.

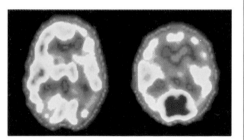

Scan comparing chemical balance of a normal brain (on left) with a schizophrenic brain.

The most common type of biological treatment is with drugs (see page 24).

Psychological treatments, or therapies, try to help people to change the conduct, beliefs and attitudes which are part of their illness, without using drugs. They usually involve the patient getting to know a therapist and working through their problems with him or her. This can take months or years.

STROKES

There are many diseases of the brain which don't cause mental illness.

Strokes, for example, are caused when a blood vessel in the brain bursts or is blocked. The cells around the vessel don't get enough oxygen so they die. Depending on where the blockage occurs, different functions are damaged. People may have problems with speech, movement or memory. One of the oddest defects is when people behave as if they can only see things on their right side, although their eyes are fine. This defect is known as left blindness. Read about it on page 56.

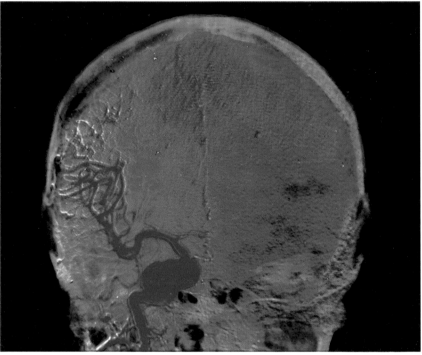

Rear view of the head showing a blockage in an artery on the left-hand side of the brain.

PARKINSON'S DISEASE

Parkinson's disease destroys the neurons that produce a chemical called dopamine. The main part of the brain that is affected by dopamine, is the part that controls movement. Patients often tremble uncontrollably, lose their balance and have problems carrying out the simplest actions like stirring a cup of tea. Drugs relieve the symptoms, but there is no cure.

DRUGS

Drugs change the balance of chemicals in your brain. They are vital in modern medicine and can save lives. But they are also extremely dangerous. They can be addictive (you can't give them up without suffering painful withdrawal symptoms), may change your personality and, if you overdose, they can kill you.

There are four main types of drugs: sedatives, painkillers, hallucinogens and stimulants.

SEDATIVES

Sedatives slow down the brain, and make you feel sleepy or calm. Doctors often give them to people who are suffering from anxiety. But it's easy to become dependent on these drugs. People begin to think they couldn't cope without them.

Alcohol is a sedative. Small amounts can make people feel relaxed and confident. Larger doses slow down

reactions so much that people start slurring their words, become unable to make intelligent decisions and lose their sense of balance.

PAINKILLERS

You've probably taken painkillers, like aspirin, when you've had a headache. Painkillers block the chemicals which make you feel pain. Morphine and heroin are very strong painkillers. They are made from opium, which grows in a kind of poppy. Morphine is given to people in severe pain. Heroin is usually taken illegally. You can read about how the brain produces its own painkillers on page 51.

HALLUCINOGENS

Hallucinogens cause hallucinations (see page 22). LSD (also called "acid") is one of the most common illegal hallucinogens. It is taken on small squares of blotting paper which are dissolved on the tongue.

Acid hallucinations may be exhilarating, but they can also be vivid nightmares. Afterward, users often feel upset. Mentally ill people can be particularly seriously damaged.

STIMULANTS

Stimulants speed up activity in your brain and make you feel more alert and sensitive to sights, sounds and feelings. They can help people suffering from severe depression (see page 22). Cocaine and crack are illegal stimulants. They may make users feel good for about 30 minutes, but later they often feel extremely tired and depressed.

OUT OF THIS WORLD?

Have you ever been thinking of someone and only moments later they've phoned you; or had a dream about something that then occurred? Many people believe that incidents like these are the result of mysterious powers of the human brain.

PSI

Psi is what scientists call the unexplained communication of information. It includes extra sensory perception, or ESP, and psychokinesis (the ability to influence events or objects using only mental powers).

There are three types of ESP:

Telepathy - transferring information from one person to another using only thoughts.

Precognition - predicting the future.

It's the Queen of Hearts.

Clairvoyance - being able to perceive things without using your senses.

Believers in psi have tried to set up experiments to prove that it exists. Most scientists remain unconvinced. They think psi has more to do with coincidence than anything else. But many people's own experiences leave them certain that some people have powers that science cannot explain.

HYPNOSIS

Hypnosis used to be thought of as a kind of black magic, which could make people perform amazing feats. Now scientists think hypnosis is a state of extreme suggestibility. Someone under hypnosis will, when commanded by a hypnotist, do things which they would not usually believe they were capable of, but they do not gain super-human powers.

What happens to a person under hypnosis:

Do you want tea or coffee?

She loses her ability to make decisions.

Her attention becomes selective - she will only hear and see what she is told to hear and see.

She may be able to go back in time to experience, for example, her 4th birthday party. Scientists aren't sure if these are true memories or just very vivid imaginings.

When instructed to do so, she can be made to forget what happened during the hypnosis. At a prearranged signal, the memories can be restored.

ANIMAL BRAINS

Now look... this really isn't difficult...

What makes humans more intelligent than any other animal? The answer is, of course, our brains. It is not simply a matter of size. Whales and elephants both have bigger brains than humans, but they would have difficulty taking a simple IQ test!

What counts more is the relative size of our brains. A human brain weighs about 1.35kg (3lbs) and makes up about 2% of human body weight. The world's biggest brain belongs to the sperm whale, which weighs in at 9kg (20lbs) - only 0.02% of body weight. The size of the human cerebrum also sets our brains apart. We have a bigger cerebrum than any other animal.

But although our brains are the most complicated, even the simplest and tiniest animal brain can still do some truly amazing things.

LEARNING

Every animal is born with instincts - things it can do automatically. Many animals survive on instincts alone. But others have the ability to learn skills, using a brain.

THE HONEYBEE

The honeybee has a tiny brain that weighs less than 0.01g (0.0004oz). Yet it has an amazing capacity to learn complex information.

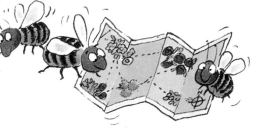

It can learn which flowers give the best pollen and at what times of the day, and all the landmarks within 1 square km (0.4 square mile) of its hive.

mmm... Yum!

Honeybees can also pass on information they have learned to other bees. When a bee has found a patch of flowers it goes back to the hive and does a dance to tell the other bees about it.

This dance is called the waggle-dance. It tells other bees how to get to a source of pollen.

NUTCRACKERS

Many animals hide stores of food for the winter. But these are useless if they can't be found again. So these animals have to have good memories. The prize for the best memory goes to a bird called Clark's nutcracker. It stores pine seeds in up to a thousand different places and can find them all again. A human couldn't manage such a feat.

For a link to a website where you can discover what different animals' brains look like, go to www.usborne-quicklinks.com

NOT-SO-SLUGGISH?

The seaslug has a brain made up of only 20,000 neurons (a human brain has billions) but it can still learn.

The seaslug is gently touched on its side and doesn't react.

A strong jet of water is then immediately squirted at it. The seaslug recoils at the water jet.

This is repeated several times.

Eventually the seaslug recoils as soon as it is touched. It has learned that one stimulus (a gentle touch) leads to the other (a jet of water).

SOCIABLE APES

Monkeys and apes have brains similar to our own. Like us, they live in large social groups and form complex relationships. Survival does not depend only on getting food and defending themselves. They also need to be able to get along with other monkeys and apes, and to know their status in the group.

A female chimpanzee finds some bananas in a clearing in the forest.

She is about to start eating, when she spots a male approaching.

She hides the bananas, and looks around innocently as if just passing the time of day.

Only when the male turns and walks away into the forest, does she start eating.

But unfortunately for her, the male has stopped behind a tree and is spying on her.

She reluctantly hands the bananas over to him and runs off into the forest.

The female chimp has used her brain to analyze her situation. She identified the male as being more dominant than her, and understood this would mean the bananas would be taken from her. Once spotted though, she knows she has to give them up to avoid being hurt in a fight. Very few animals can manage this kind of thought.

COMPUTER BRAINS

Will the computer ever become more intelligent than the human brain? It's already happened in fiction. In books and films, computers often not only have vast stores of knowledge, but they also have morals, great personalities and can tell jokes. In reality, so little is known about how the brain works, the idea of a computer being able to imitate or better it is unimaginable.

This is Artoo Deetoo, the intelligent computer from the films *Star Wars* and *The Empire Strikes Back*.

INTELLIGENT MACHINES?

In some areas computers are much more efficient than humans. They can analyze huge amounts of data and do long, complicated calculations in a fraction of a second. They can beat all but the very best players at chess. They can help doctors diagnose diseases. Computerized robots can do the work of highly skilled mechanics.

We think of some of these skills as signs of great intelligence in people. But this does not mean that computers are intelligent. In fact, all the computer is doing is following a set of rules which was programmed into it by an intelligent person.

These robots can put together the fiddliest parts of cars without taking a break or losing concentration.

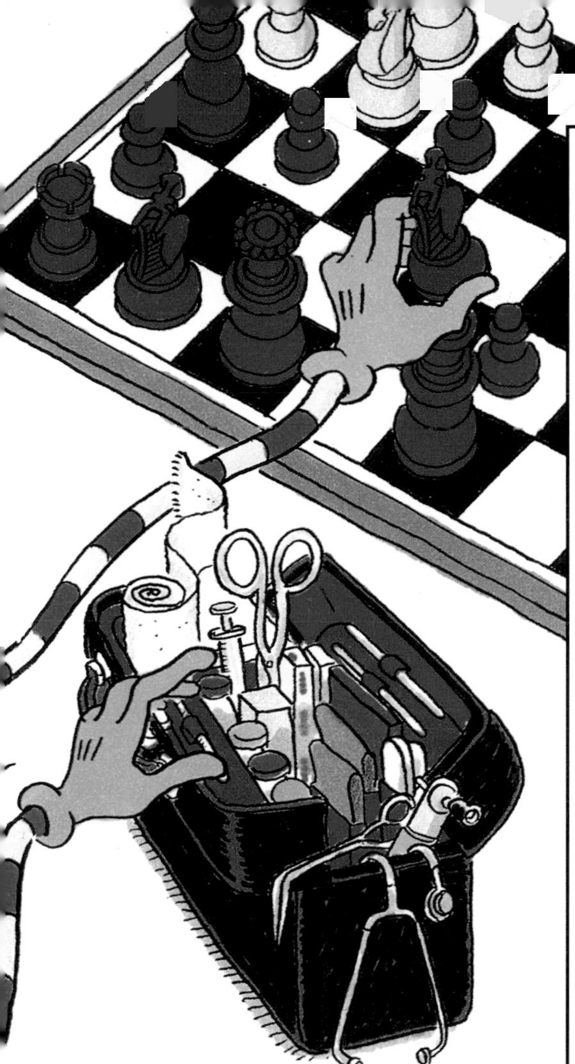

Read this list of words.

animal

animal

animal

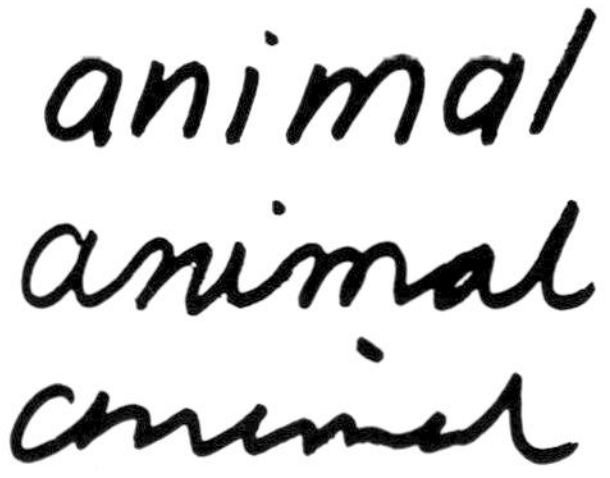

That was easy, wasn't it? Each word says animal, but in different handwriting. The last word is practically illegible, but because of the words above it, you can guess what it is.

A computer programmed to read handwriting would fail on at least one of these words. This is because computers can follow instructions but are bad at taking a guess.

HUMANIZING COMPUTERS

Although it's false to think of computers as being intelligent, it is true to say that scientists are making computers that behave in more and more human ways. This makes computers easier to use and able to perform functions that are more useful to humans.

One group of people who can benefit enormously from the humanizing of computers is the disabled. The more "human" computers become, the more they will be able to help people.

Computers have given the scientist Stephen Hawking the "freedom" to carry on his work despite being disabled.

There is even a possibility of developing a computer that can act like a part of your nervous system (see page 7). This could give people whose spinal cords have been severed a chance to walk again.

This diagram shows how a computer may one day help someone whose spinal cord is broken to walk.

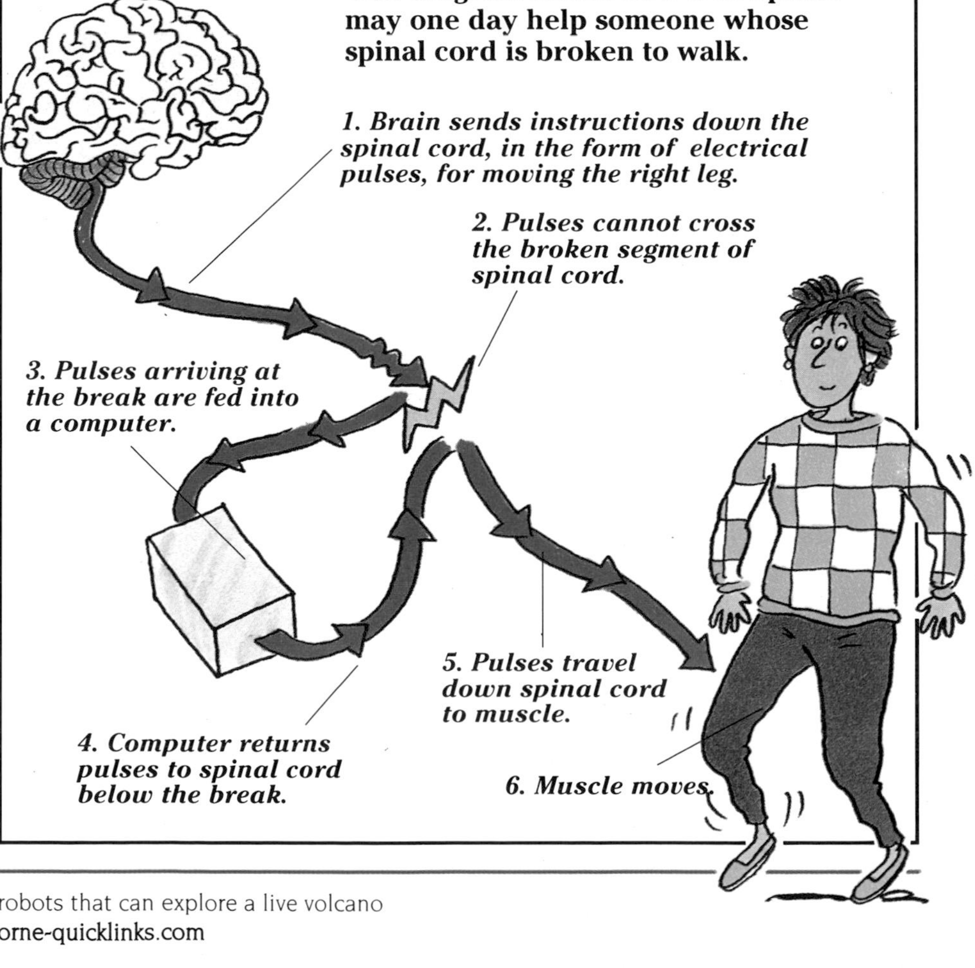

1. Brain sends instructions down the spinal cord, in the form of electrical pulses, for moving the right leg.

2. Pulses cannot cross the broken segment of spinal cord.

3. Pulses arriving at the break are fed into a computer.

4. Computer returns pulses to spinal cord below the break.

5. Pulses travel down spinal cord to muscle.

6. Muscle moves.

BRAINS IN HISTORY

Throughout the ages the brain has been a riddle to scientists. Even today, scientists only understand a fraction of what goes on inside your head.

ANCIENT IDEAS

The Ancient Greeks were some of the world's first scientists. They explored many areas of science, including what happens in the human body.

They had many different theories about where in the body thoughts, feelings and emotions came from.

The poet Homer, who lived about three thousand years ago, thought they came from the lungs.

The great Greek scientist, Aristotle (384-322BC), believed that they came from the heart. Many of us still think love comes from the heart, although we know that this is nonsense according to science.

The first really scientific look at the brain was made in the third century BC, by the Greek scientists Herophilus and Erasistratus.

They were some of the first people to dissect (cut up) animal and human bodies in order to find out about what goes on inside.

Their most important work was the discovery of the nervous system (see page 8). This showed that the brain was in charge of much of what happens in the body.

Galen, doctor to the Roman emperors in the 2nd century AD, continued this study of the brain and nervous system. But, because he did much of his research on animals, not humans, he got some things completely wrong. However, he was still considered the world's brain expert for over a thousand years.

For a link to a website where you can follow a day in the life of a child's brain, go to www.usborne-quicklinks.com

PHRENOLOGY

In Europe and America, from the middle of the 18th to the middle of the 19th century, phrenology was a very popular brain science. Phrenologists thought they could analyze a person's character from the shape and bumps of their skull.

They believed the skull was shaped by the structure of the brain beneath it, and that different parts of the brain were responsible for very specific characteristics, skills and talents.

The size of the temples, (the area above the cheekbones), for example, were meant to reveal how musical someone was, while the shape of the base of their skull was meant to show if someone would be a good parent.

Phrenology was taken so seriously that, for a time, it was used to select people for jobs. There was even a suggestion that children's heads should be shaped to bring out good characteristics, and to suppress bad ones.

LOCALIZATION

The idea that different parts of the brain have different functions is called localization. Phrenology was the silliest of all localization theories. But there were serious scientists in the late 19th century who studied localization. Broca and Wernicke both studied the brains of dead stroke patients and discovered the part of the brain that controlled language.

But it was not really until the 20th century, that the mysteries of what actually happens inside the brain were begun to be resolved. With the arrival of new technology, such as brain scans, better microscopes and advanced brain surgery, doctors and scientists have been able to look closely at the brains of living people.

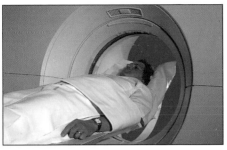

A patient having a brain scan

As more data is collected, more can be understood about what your brain can do. Much, however, remains a mystery.

TREPANNING

Trepanning was an ancient medical practice, that survived until the Middle Ages. It involved making a small hole in someone's skull. This was meant to release evil spirits that were making the patient insane. It may actually have helped some patients with brain growths, but in most cases it probably did more harm than good.

A GLOSSARY

Here's a reminder of what some of the difficult words in this book mean.

Cerebral hemispheres – The two domes into which the top of your brain is divided.

Consciousness – What you are aware of at any one moment.

Craniologist – A scientist who studies the shape and size of the human skull.

Genes – Tiny chains of chemicals in your cells. They contain instructions which control how your body works.

Homeostasis – Keeping your body and the chemicals inside it in a stable state.

Hormones – Chemicals released by organs in your body that carry instructions to your cells telling them what to do.

Hypothalamus – A tiny part of your brain that controls your heart rate, waterworks, temperature, sleep and sexual development.

Neurologist – A scientist who studies the brain and nerves.

Nervous system – A network of nerve cells that stretches from your brain to the tip of your toes.

Neurons – The cells which make up your brain and your nerves.

Phrenologists – People who believe they can analyze a person's character from the shape of their skull.

Psi – Communicating information without using the senses, for example using only thoughts.

Psychiatrist – A doctor who treats people when their brains don't work right and they act in strange ways.

Psychologist – A scientist who studies how humans think and behave.

REM – A period of sleep when your brain is very active. This is the time when you dream.

Retinal image – The picture formed on the back of your eye of the things you are looking at.

Schizophrenia – An illness in which people may hear voices that aren't there, or begin to believe things that aren't true.

Synapse – A tiny gap between one nerve cell and another.

DID YOU KNOW?

A dinosaur called a Stegosaurus lived about 150 million years ago. It grew up to 9m (30ft) long, but had a brain the size of a ping pong ball.

Monkeys and apes are probably the most intelligent animals. Johnnie, a monkey in Australia, learned how to drive a tractor. Johnnie was also able to understand commands such as "turn left", and "turn right".

In 1810, a boy named Zerah Colburn from Vermont, in the United States, multiplied 12,225 by 1,223. He was only six years old at the time.

In 1996, an American man named Dave Farrow tried to memorize 2,704 playing cards. When he was asked to remember the cards, and the order in which they appeared, he only made six mistakes.

Understanding your
Senses

INTRODUCING SENSES

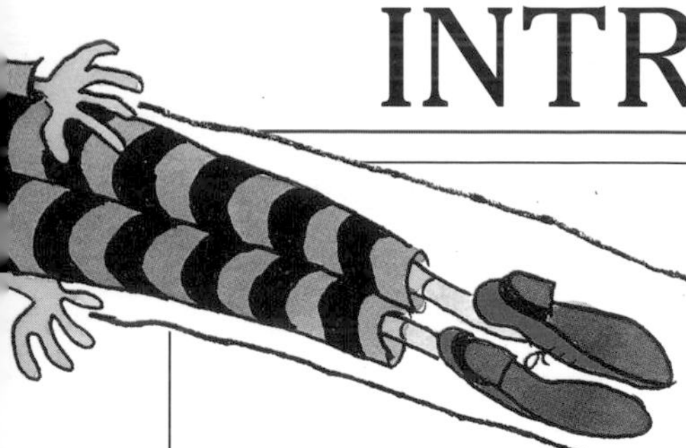

This part of the book is all about senses. Your senses let you know what is happening in the outside world and to your own body. Without them, you would be completely cut off from everything around you. You wouldn't even know whether your arms were folded or your legs crossed. You have five main senses: sight, hearing, taste, smell and feeling, or touch. The parts of your body which can sense things are called sense organs.

CONTENTS

34 Introducing senses

36 Getting information

38 Eyes & seeing

40 More than meets the eye

42 Ears & hearing

44 Blindness & deafness

46 Taste & smell

48 Feeling touchy

50 A painful subject

52 Tricky positions

54 Brain blindness

56 Animal senses

58 Learning to sense

60 Just an illusion?

62 Smart machines

64 A glossary

You need your senses to do all the things shown in these pictures.

Know the position of parts of your body

Touch or feel

See

Balance

GETTING THE PICTURE

Sight is the sense which gives you the most information about the world. But if you were blindfolded, you'd be surprised how quickly your other senses could help you find out about what is around you.

Hands feel objects around you.

Ears tell you where people are.

Smell leads you to food.

Taste tells you it's good to eat.

Smell

Taste

Hear

SIXTH SENSE

Sometimes you can have a vague "feeling" that something is happening, without seeming to see, hear, taste, smell or feel it. People often call this their "sixth sense". In fact, your senses have probably picked up something so tiny that you weren't even aware of it, such as a tiny breath of wind.

Hello, I'm Norman the Neuron.

You will find out how important neurons, or nerve cells, are on page 36.

GETTING INFORMATION

The parts of your sense organs that can tell what is happening are called receptors. Receptors convert information into electrical pulses which are sent to your brain along nerve cells. Your brain analyzes the information and makes you aware of what is going on. It may then send instructions back to parts of your body, making them act upon this information.

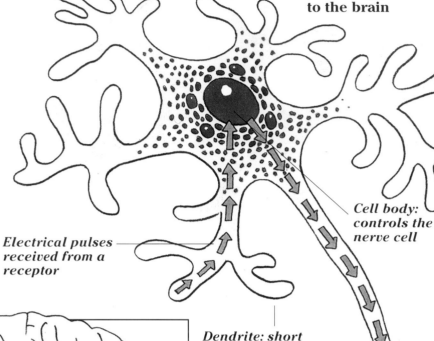

Nerve cells sending pulses to the brain

Electrical pulses received from a receptor

Cell body: controls the nerve cell

Dendrite: short tentacle which carries pulses to the cell body

Axon: long tentacle which carries pulses away from the cell body

NERVE CELLS

Nerve cells carry pulses of electricity from receptors to your brain. A nerve cell has a cell body, an axon, and tentacles called dendrites. The pulses pass from the end of an axon to the nearest dendrite of the next cell. The pulses are passed from one cell to the next, until the information reaches the brain.

You can read more about nerve cells and how the pulses pass between them on pages 6 and 7.

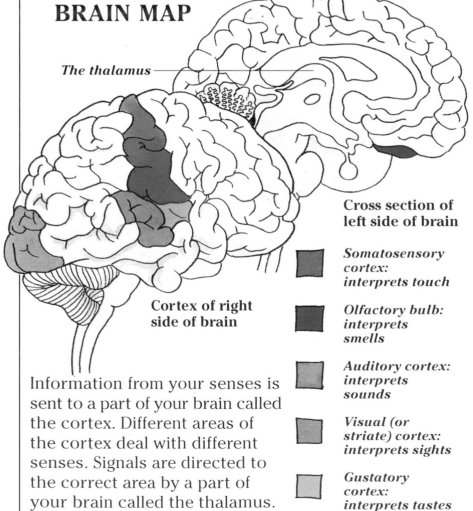

BRAIN MAP

The thalamus

Cross section of left side of brain

Cortex of right side of brain

Somatosensory cortex: interprets touch

Olfactory bulb: interprets smells

Auditory cortex: interprets sounds

Visual (or striate) cortex: interprets sights

Gustatory cortex: interprets tastes

Information from your senses is sent to a part of your brain called the cortex. Different areas of the cortex deal with different senses. Signals are directed to the correct area by a part of your brain called the thalamus.

DIFFERENT RECEPTORS

The receptors in your different sense organs are designed to detect and respond to different things. This is why you cannot see with your ears or smell with your eyes.

Some receptors in your ears respond to sounds. Others respond when you turn your head so you can keep your balance.

Receptors inside your nose respond to chemicals in the air.

Receptors in your tongue respond to liquids or substances dissolved in saliva.

Receptors in your skin respond to touch, pressure, temperature and pain.

Receptors in your eyes respond to light and color.

Receptors in your muscles and joints respond when you move, so you know the position of parts of your body.

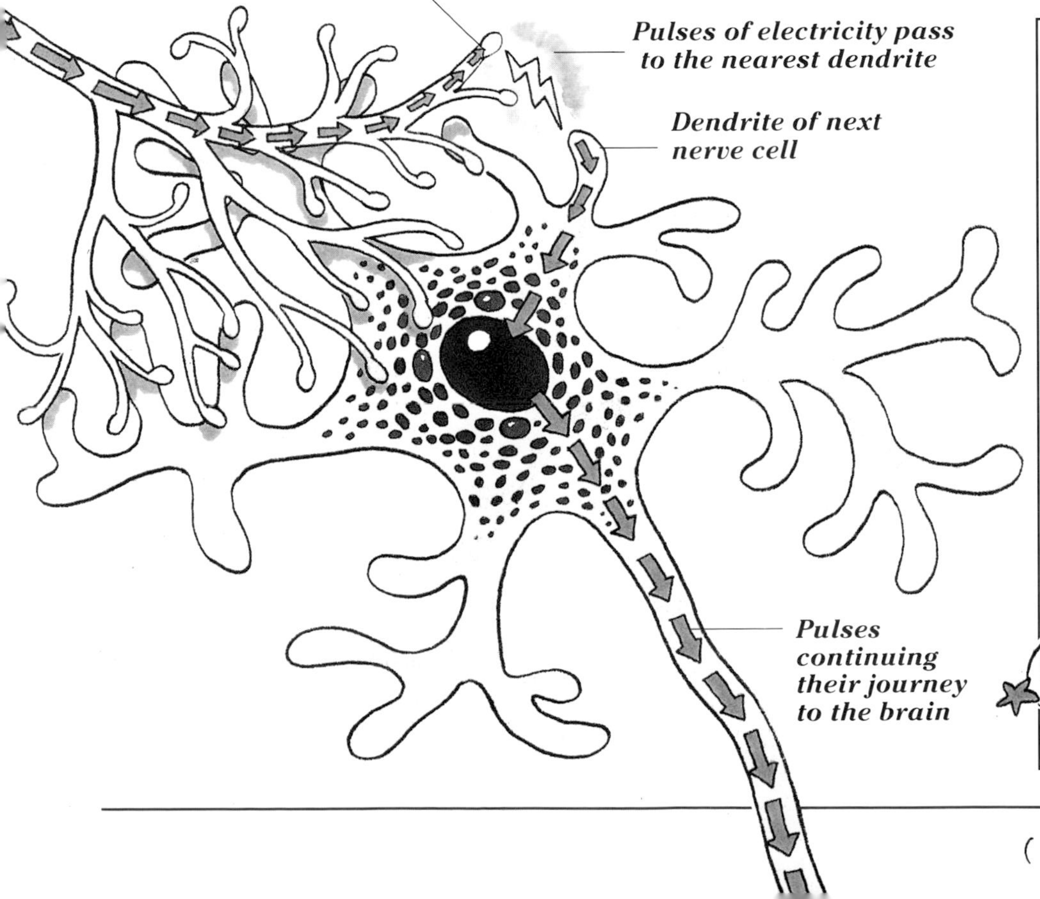

End of axon

Pulses of electricity pass to the nearest dendrite

Dendrite of next nerve cell

Pulses continuing their journey to the brain

SEEING STARS

Receptors in your eyes are only meant to respond to light. But a bad bang on the head can make them fire off signals to your brain. As the signals come from your eyes, the thalamus sends them to your visual cortex. This means you see flashing lights or stars.

EYES & SEEING

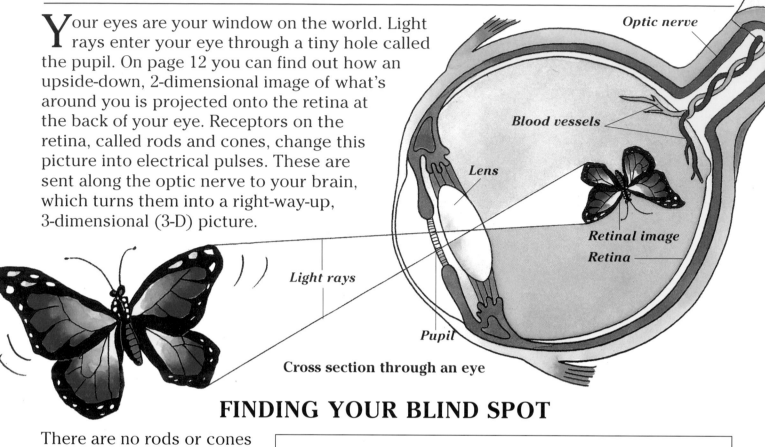

Your eyes are your window on the world. Light rays enter your eye through a tiny hole called the pupil. On page 12 you can find out how an upside-down, 2-dimensional image of what's around you is projected onto the retina at the back of your eye. Receptors on the retina, called rods and cones, change this picture into electrical pulses. These are sent along the optic nerve to your brain, which turns them into a right-way-up, 3-dimensional (3-D) picture.

Optic nerve

Blood vessels

Lens

Retinal image

Retina

Light rays

Pupil

Cross section through an eye

FINDING YOUR BLIND SPOT

There are no rods or cones on the tiny spot where the optic nerve leaves the eye. This is called your blind spot. If an image falls exactly on your blind spot, you can't see it. You don't usually notice your blind spot because images usually fall on other parts of your retina too, and your brain just fills in the gap.

Line up the square man with your left eye. Close your right eye. Move the book slowly **away from you. When the book is about 30cm (1ft) away, the round man disappears.**

GETTING THINGS IN FOCUS

The lens is short and fat to focus on a nearby object.

The lens is longer and thinner to focus on a distant object.

A lens in your eye makes sure the objects you look at are in focus (not blurry). It changes shape when you look at things at different distances. This makes the rays entering your eye bend by different amounts.

WHY HAVE TWO EYES?

Hold a finger up about 25cm (10in) from your nose. Then focus on an object in the distance behind it. Shut each eye alternately, and watch your finger jump from side to side.

This happens because each eye sees your finger from a slightly different angle. Your brain joins the two images to help you see in 3-D. This is called binocular vision.

A RAINBOW WORLD

Most colors can be made by mixing the three basic colors of light: red, blue and green.

You can 'mix' colors in your eye in the same way. You have three types of cones on your retina: red, blue and green. Each type responds a different amount depending on what color you are looking at.

If you are looking at purple grapes, the blue and red cones respond more strongly than the green cones.

COLOR-BLINDNESS

Normal color vision   *Colorblind vision*

Colorblind people can't tell the difference between red and green, because their red or green cones do not work properly.

They learn to call grass green, for example, just because other people do. Occasionally, people have normal vision in one eye, but are colorblind in the other one.

SEEING IN THE DARK

Cones don't work very well in dim light, so this is when you use your rods to see. Rods can't detect color, which is why it is hard to make out colors at night. But even rods need some light to work. They are helped by your pupils, which grow bigger in the dark. This is to let in as much light as possible.

While you are out in bright sunlight, you use your cones to see. Your rods don't work.

If you suddenly walk into a dark place, you can hardly see anything at first.

Your pupils soon enlarge, so more light enters your eyes. Your rods start working.

You may not like what you see lurking in the darkness!

For a link to a website where you can explore an online exhibit and find out how your mind sees color, go to **www.usborne-quicklinks.com**

MORE THAN MEETS THE EYE

A green tennis ball always looks like a green tennis ball, whether it's flying through the air in sunlight, or in the corner of a dark cupboard. This is because your brain uses four 'constancy mechanisms': size, brightness, color and shape. This means that you can recognize the things you look at, even though their image on your retina can be completely different in different situations.

SIZE

Two things that are the same size will project retinal images* of different sizes if one is farther away than the other. But your brain knows they are the same size.

*See picture of eye on page 38.

Your brain knows these ice cream cones are the same size, although their retinal images are different.

SHAPE

Shape constancy means that your brain tells you an object is the shape it would be if you were looking at it from straight on, even when you look at it from a different angle. So you know a door is a rectangle, even if the image on your retina is not rectangular at all.

BRIGHTNESS

Light things are brighter than dull things. But a light thing in a dim room can actually be duller than a dull thing in bright sunlight. But your brain still sees the thing as light, because it compares it to other duller things in the room.

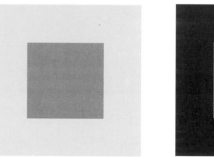

Because your brain compares things to their surroundings, this can lead to illusions. These gray squares are identical, but they don't look like it. The lighter the background, the darker the gray squares appear.

COLOR

Color constancy means a green apple still looks green, even if you are looking at it through red glasses. This is because it is still greener than everything else around it.

Perhaps I'll have a red apple after all!

HOW YOU SEE IN 3-D

Your brain is very good at figuring out how far away things are. It uses visual signals, called cues, in the retinal image to determine depth. This, along with binocular vision, gives you a 3-D picture of what you see.

Like a retinal image, this picture has cues to help you see it in 3-D.

Your brain interprets arrow-shaped lines as corners.

This tree overlaps the next tree. This tells you it is nearer.

This train is smaller than the other. This tells you it is farther away.

Your brain knows that lines which appear to get closer together (like these railroad tracks) are in fact parallel lines getting farther away.

SEEING WHAT YOU WANT TO SEE

Often what you see depends on what you are expecting to see, or what you want to see. For example, if you are hungry, you might catch a fleeting glimpse of a red ball and think it is a tomato. Or, if a familiar word is misspelled, you may not notice. This is because your brian has just assumed it is right. (Did you spot the deliberate mistake in this paragraph?)

The things around the object you look at also affect the way your brain analyzes it.

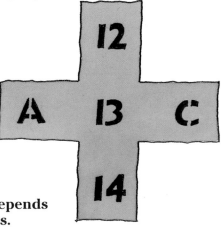

How you see the middle symbol depends on which way you look at the cross.

EARS & HEARING

I f you drop a pebble into the middle of a pond, waves or ripples travel outward to the edges of the water. Sounds make the same sort of waves in the air, but of course you can't see sound waves. When sound waves travel into your ear, you can hear them.

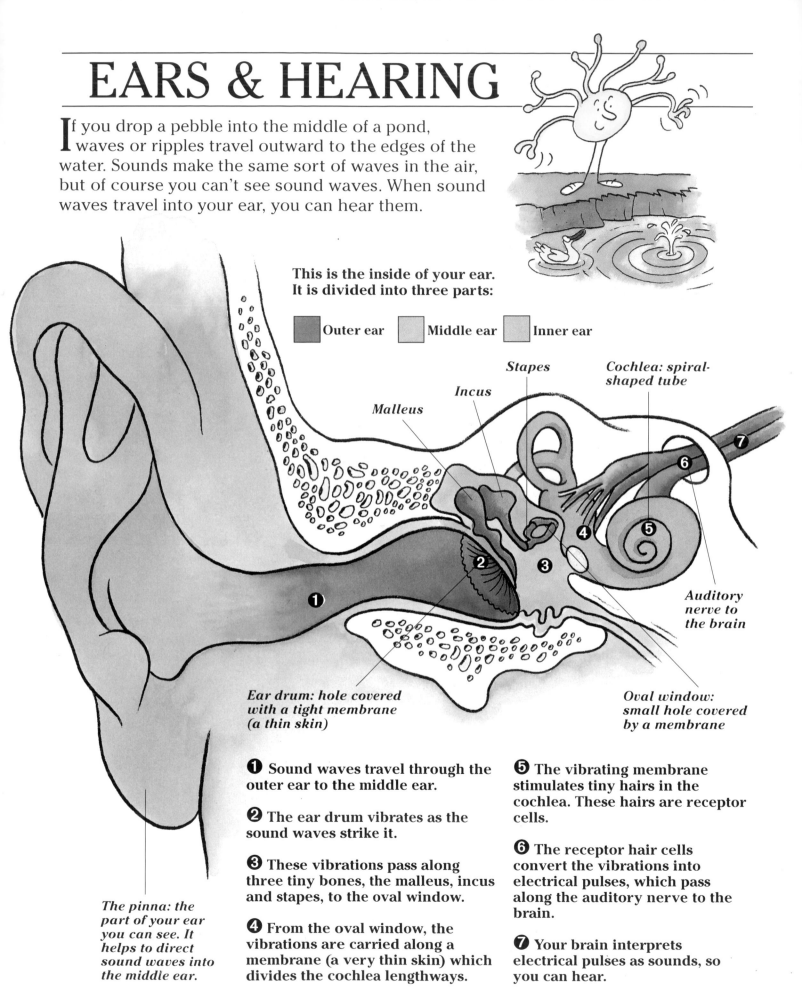

This is the inside of your ear. It is divided into three parts:

Outer ear **Middle ear** **Inner ear**

Malleus

Incus

Stapes

Cochlea: spiral-shaped tube

Auditory nerve to the brain

Ear drum: hole covered with a tight membrane (a thin skin)

Oval window: small hole covered by a membrane

The pinna: the part of your ear you can see. It helps to direct sound waves into the middle ear.

❶ Sound waves travel through the outer ear to the middle ear.

❷ The ear drum vibrates as the sound waves strike it.

❸ These vibrations pass along three tiny bones, the malleus, incus and stapes, to the oval window.

❹ From the oval window, the vibrations are carried along a membrane (a very thin skin) which divides the cochlea lengthways.

❺ The vibrating membrane stimulates tiny hairs in the cochlea. These hairs are receptor cells.

❻ The receptor hair cells convert the vibrations into electrical pulses, which pass along the auditory nerve to the brain.

❼ Your brain interprets electrical pulses as sounds, so you can hear.

AMPLITUDE AND PITCH

Sounds have different amplitude and pitch. Amplitude is another word for loudness. Pitch is how high or low a sound is. As a sound travels, its amplitude gets smaller. So the further you are away from the source of a sound, the quieter it is.

Sounds of different amplitudes make waves of different heights. Sound A is louder than B.

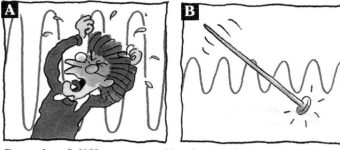

Sounds of different pitch have different frequencies. Frequency is the number of waves passing a given point in one second. Sound C is higher than D.

A violin, piano and saxophone all playing the same note at the same loudness don't all sound the same. This is because the sounds they produce have higher but quieter notes mixed in. These are called harmonics. Harmonics give an instrument its own individual sound.

DECIBELS

Loudness is measured in decibels (dB). The softest sound anyone can hear is 0dB.

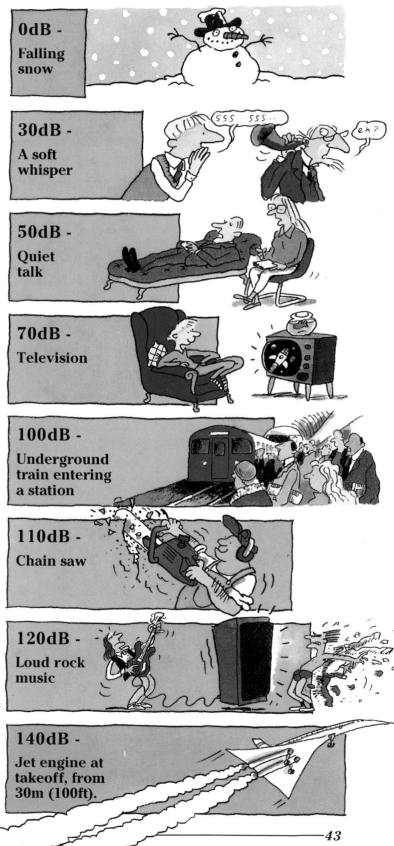

0dB - Falling snow

30dB - A soft whisper

50dB - Quiet talk

70dB - Television

100dB - Underground train entering a station

110dB - Chain saw

120dB - Loud rock music

140dB - Jet engine at takeoff, from 30m (100ft).

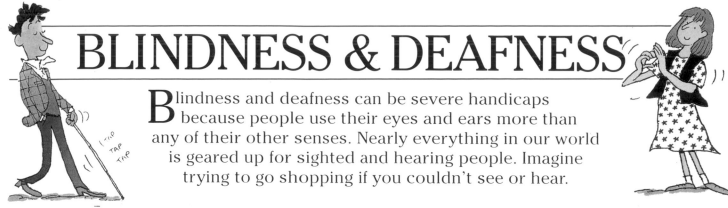

BLINDNESS & DEAFNESS

Blindness and deafness can be severe handicaps because people use their eyes and ears more than any of their other senses. Nearly everything in our world is geared up for sighted and hearing people. Imagine trying to go shopping if you couldn't see or hear.

BLINDNESS

Severe damage to any part of the eye can cause blindness. Some people are born blind; others become blind because of disease or an accident.

Very few people can see nothing at all. These pictures show what a fruit stall looks like to people with different kinds of blindness.

Macular degeneration – damages the spot on the retina which lets you see best. A big blindspot develops.

Severe glaucoma – a disease of the eye which damages nerve cells. Can lead to "tunnel vision".

Diabetic retinopathy – damage to the retina, caused by diabetes. Sight becomes patchy and blurred.

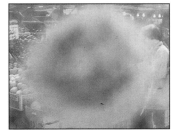

Cataracts – make the lens go cloudy. Can happen when people get older, or because of injury to the eye.

There are no cures for most types of blindness, but cataracts can be cured by a simple operation. Sadly, in many countries, there is not enough money to perform the operation on everyone who needs it.

DEAFNESS

There are lots of causes of hearing problems. Some lead to only a small amount of deafness, but others can leave you totally deaf.

Some babies are born with hearing problems. Sometimes an illness such as meningitis can make you go deaf. As people get older, often their ears just stop working so well. Loud noises can also damage ears and cause deafness.

Ear protectors can prevent deafness.

HEARING AIDS

Many people who find it hard to hear wear hearing aids. These make sounds louder.

Hearing aids are made up of these basic parts:

Microphone – receives sound waves and turns them into electrical signals.

Amplifier – increases the strength of the electrical signal.

Earphone – converts electrical signals from the amplifier back into sound waves.

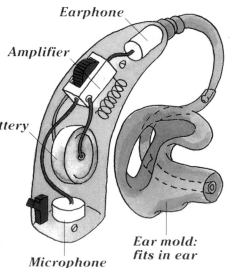

Earphone

Amplifier

Battery

Microphone

Ear mold: fits in ear

LIP READING

In a noisy place, hearing aids make everything louder so wearers often can't hear their own conversations. So, for many deaf people, lip reading is easier. Lip readers figure out what people are saying by looking at their lips, tongue and neck muscles. But even the best lip readers read only about half of the words spoken. They have to guess the rest.

Back to the light   *Hand on mouth*   *Beard over lips*

The things above make it hard for people to lip read.

SIGN LANGUAGES

People with serious hearing problems often use sign languages to talk to each other. Some sign languages use one hand gesture for each letter of the alphabet. Others use gestures, signs and facial expressions to express whole words, phrases and emotions. These sign languages are often incredibly complex, and have their own grammar and vocabulary.

"Your"   *"name"*   *"what?"*

This boy is asking "What is your name?", in British Sign Language (BSL). Try to answer using this alphabet sign language.

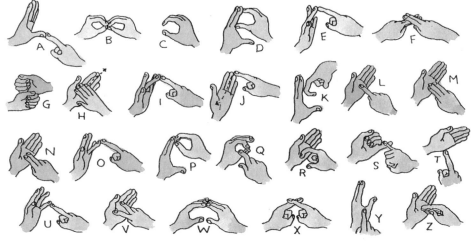

'P' OR 'B'?

People who are born deaf speak differently from hearing people. This is because they have never heard speech. Most of us learn to speak by imitating what we hear. A deaf person learns to imitate a sound or word by how it looks and feels.

**1. Look in a mirror and say 'p'. Now say 'b'.
Your lips moved in the same way for both sounds.**

2. Now say them again, but put your hand in front of your mouth. Can you feel the difference in the amount of air escaping?

This is how a deaf child learns to speak.

TASTE & SMELL

You eat and drink because you need energy to live, and energy comes from food. But it would be really boring if your food had no taste or smell. Your receptors allow you to recognize a whole range of tastes, from chocolate to chili con carne.

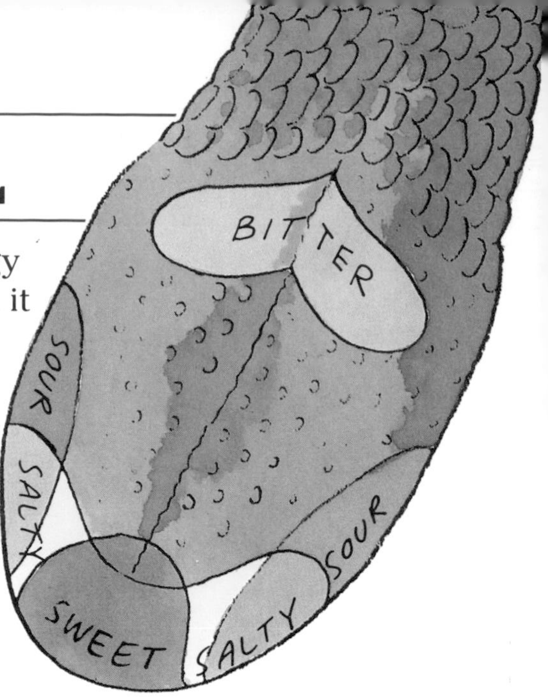

TONGUE-TASTIC

Scientists think that all flavors are made up of four basic tastes: sweet, salty, sour and bitter. The things you eat are a mixture of these tastes. So, for example, oranges are sweet and sour, grapefruits less sweet and more sour, and chips are salty and a bit sweet.

Receptors on your tongue respond to chemicals in your food dissolved in saliva. Different parts of your tongue respond to each of the four basic tastes.

The picture above shows which parts of the tongue respond to which tastes.

Section of tongue

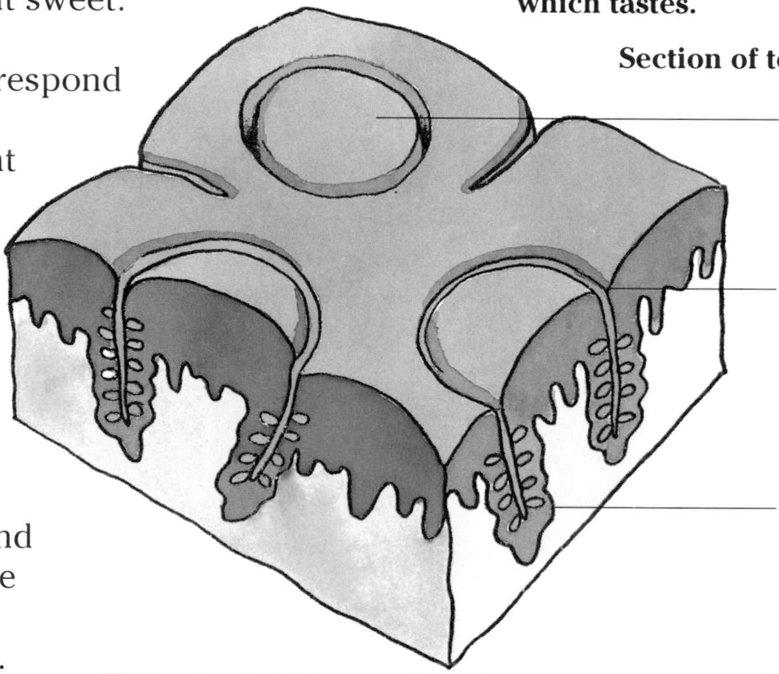

Your tongue contains hundreds of small bumps.

Each bump is surrounded by a small trench, which traps saliva.

Taste receptors lie in these trenches.

YUM OR YUK

The main purpose of your sense of taste is to tell you whether something is safe to eat. Dirt, muddy water and most poisonous plants taste horrid. So your immediate reaction is to spit them out. Most foods that are good for us don't taste nasty.

SWEET TOOTH

Many of us have a sweet tooth. This is because thousands of years ago, sweet things were extremely rare. But they were also extremely important, as they gave people a much needed energy boost.

So our ancestors developed a "taste" for sweet things, to make sure they would eat them whenever they found them.

SUGAR AND SPICE

If you find it hard to believe that all the flavors of all the different foods come from only four tastes, you'd be right. This is because flavors aren't only made up of tastes, but also of smells.

You use these parts of your body to smell.

This is why, if your nose is blocked up with a really bad cold, you can feel as if you are eating cotton or cardboard.

The difference between taste and smell is that taste receptors respond to dissolved chemicals, while smell receptors respond to chemicals in the air.

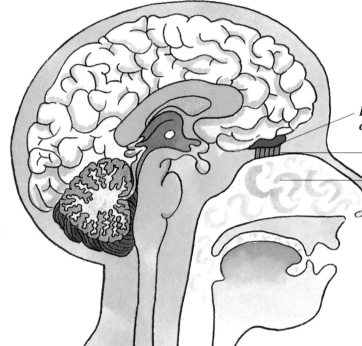

Part of brain which analyzes smell

Smell receptors

The inside of your nose is linked to your mouth, so you can smell food which is inside your mouth.

Scientists don't understand exactly how smell receptors work. But, like tastes, they think there are probably only four basic smells: fragrant (like roses), fresh (like pine), spicy (like cinnamon) and putrid (like rotten eggs).

THE SMELLY T-SHIRT TEST

In order to test people's sense of smell, a group of men and women were asked to wear the same T-shirt for 24 hours without washing. The T-shirts were then sealed in plastic bags. Each person was asked to take a sniff of three bags: one containing their own T-shirt, and two belonging to strangers – one man's and one woman's. About 75% of people could identify their own T-shirt and tell the difference between the man's and the woman's T-shirts.

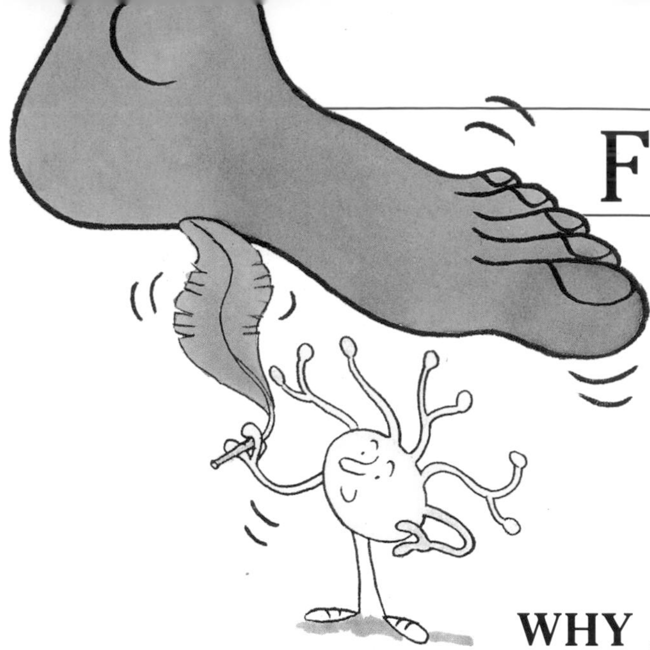

FEELING TOUCHY

The skin is not just a bag to keep your body parts in – it's the biggest sense organ in your body. It contains millions of receptors which respond to touch, pain, pressure and heat and cold. These receptors send electrical signals to your brain and give you a mass of information about what different things feel like.

WHY ARE FEET TICKLISH?

If you want to make a ticklish person squirm, tickle their feet. The soles of the feet are very sensitive to light touch, which makes them more ticklish than most other parts of your body.

In the same way that feet are more sensitive to light touch, other parts of your body are more sensitive to other qualities. Some respond most to heat, some to cold, and others to pain. Exactly what your sense of touch tells you about a particular thing depends upon the number and type of receptors in the part of the skin that touches it.

This diagram shows the main receptors found in different parts of your skin.

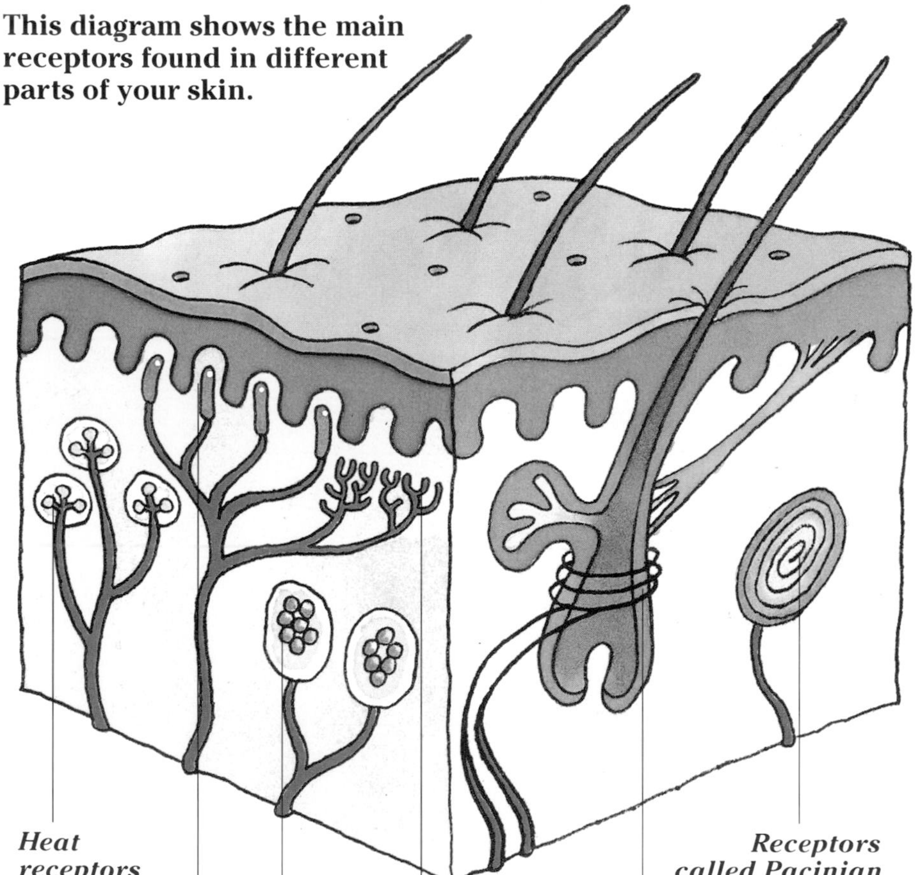

Heat receptors

Touch receptors sense light pressure. They tell you about the texture of things.

Cold receptors

Receptors called free nerve endings react to pain.

Receptors called Pacinian receptors respond to heavy pressure.

Receptors at hair roots detect pressure through the movement of the hair.

READING WITH FINGERS

Blind people use their sense of touch to read a code called Braille, made from raised dots.

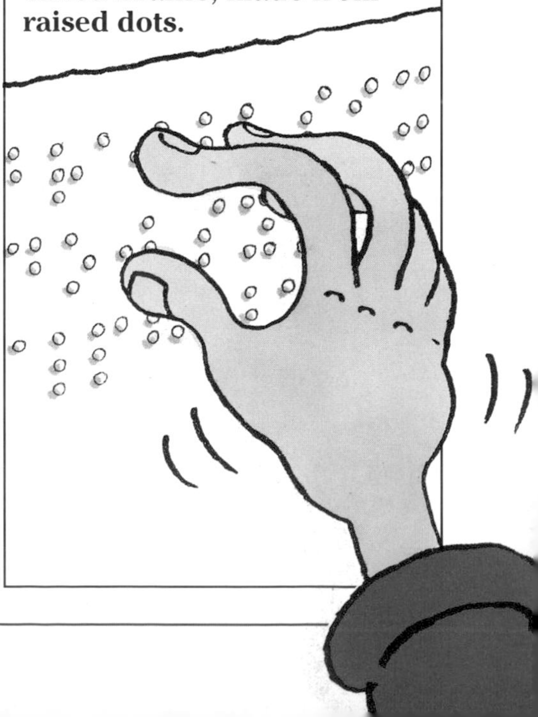

For a link to a website where you can learn the alphabet in Braille, see your name in Braille and solve some Braille riddles, go to **www.usborne-quicklinks.com**

WHAT'S THE POINT?

Your fingertips contain more touch receptors than any other part of your body. This makes them very sensitive. Try this experiment to feel the difference between the sensitivity of your fingertip and the back of your leg.

Place the points of a divider 2mm apart.

Touch a fingertip lightly. You should feel two separate points.

Now touch the back of your leg. It feels like just one point.

HOT AND COLD

Your skin contains heat and cold receptors. They both respond most strongly when they are first activated. If you are outside sunbathing on a hot day, and then go indoors, it feels very cold at first. But, after a while, it starts to feel normal, as your cold receptors adapt to the new situation.

BARELY THERE

You hardly notice the clothes you are wearing, even though they are touching you all the time. This is because when you are in contact with the same thing for some time, your touch receptors stop responding as they get used to it. So even clothes with annoying, itchy labels will start to feel less itchy after a while.

TESTING THE WATER

Do this experiment and feel your temperature receptors in action.

1. Prepare three bowls, one filled with cold water, one with warm water, and the third with hot water (not *too* hot).

2. Put your right foot in hot water and your left foot in cold water.

3. Then plunge both feet into the bowl of warm water.

4. Your right foot feels cold in the warm bucket, but your left foot feels hot.

This experiment works because cold receptors in the right foot are activated for the first time, so they respond more strongly than the heat receptors. But the opposite happens in your left foot, where the heat receptors respond more strongly than the cold receptors.

FEEL THE PRESSURE

When you use a pencil to write with, the pencil acts as an extension of your finger. Your pressure receptors respond to the object you touch with the pencil, even though your skin is not touching it directly. Try using a stick to touch different kinds of surfaces. Can you feel the difference?

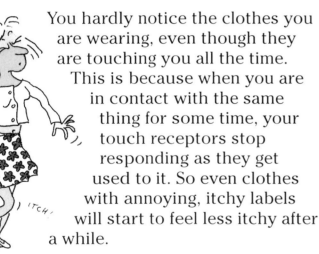

A PAINFUL SUBJECT

If someone punched you on the nose, or if you stood on broken glass, you would feel pain. Although pain is not very nice, it's very important. It tells you that your body is in danger. You feel pain when receptors respond to something which is causing you damage. They send urgent messages to your brain. You can then do something to try to stop the pain.

REFLEX ACTIONS

Sometimes, you move away from a thing that is hurting you before you actually feel the pain and know what you are doing. This is called a reflex action.

Reflex actions happen when messages from pain receptors are sent to the spinal cord, not the brain. You don't become aware of the pain, or the movement away from it, until messages are sent from the spinal cord to the brain telling it what has been happening.

This picture shows what happens during a reflex action.

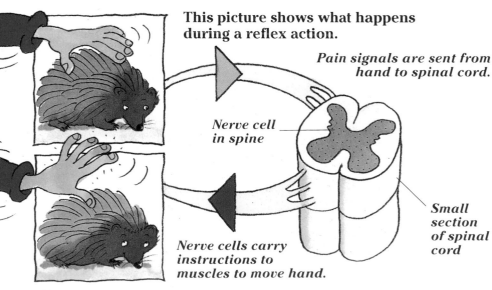

Pain signals are sent from hand to spinal cord.

Nerve cell in spine

Small section of spinal cord

Nerve cells carry instructions to muscles to move hand.

REFERRED PAIN

If someone damages an organ in their body, they may feel pain somewhere else completely. This is called referred pain.

Referred pain happens because pain receptors from different areas of your body meet at the same place in the spinal cord. The brain confuses where the messages are coming from.

This picture shows the areas where you can feel referred pain when certain organs are damaged.

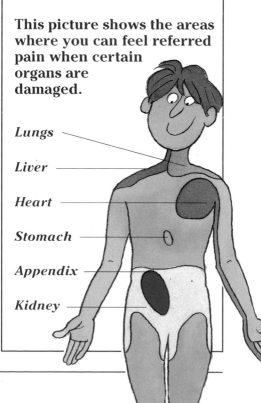

Lungs

Liver

Heart

Stomach

Appendix

Kidney

For a link to a website where you can test your reflexes, go to
www.usborne-quicklinks.com

CONTROLLING AND RELIEVING PAIN

There are lots of different ways of controlling pain. One way is by taking painkilling drugs.

Painkillers don't cure illnesses and injuries, but they are chemicals which stop you from feeling pain. One of the strongest painkillers is morphine. It is used to help control very severe pain.

Your brain also produces its own painkillers, called endorphins. These work in the same way as morphine does. They are usually released after pain is first felt, but they often stop working once the immediate crisis is over.

Endorphins mean that even people who are very badly wounded often don't feel pain from their wounds until later. This may give them the time to get help.

A man bitten by a lion feels pain.

But endorphins reduce the pain, enabling him to crawl to safety.

Soothing music can help people block out pain.

As well as producing endorphins, scientists think your brain can also block pain signals coming from the spinal cord. No one understands exactly how this works.

People seem to be able to develop their own ways of relieving pain. This might be by singing, listening to music, deep breathing, or tightly holding or biting something. Scientists can't explain why these things help.

MIND OVER MATTER

How much pain people feel is not only to do with the strength of the signals from pain receptors. Other things, such as a person's culture and attitude of mind, also have a big effect. In some Eastern religions, people take part in rituals which would be unbearably painful to Westerners. But they don't seem to feel any pain.

This man looks quite relaxed on his bed of nails.

TRICKY POSITIONS

Your brain receives messages from receptors in muscles all over your body. They tell your brain the position of all your limbs and joints. Your brain analyzes the information it is sent and uses it to coordinate all your movements. This means you can cross your legs and scratch your nose at the same time without falling over.

MOVING MUSCLES

You move parts of your body using pairs of muscles. These are attached to your bones by tough bands called tendons. By monitoring muscles and tendons, your brain works out what you are doing with your arms and legs. Find out more about muscles in the third part of this book.

These pictures show how a pair of muscles moves your arm.

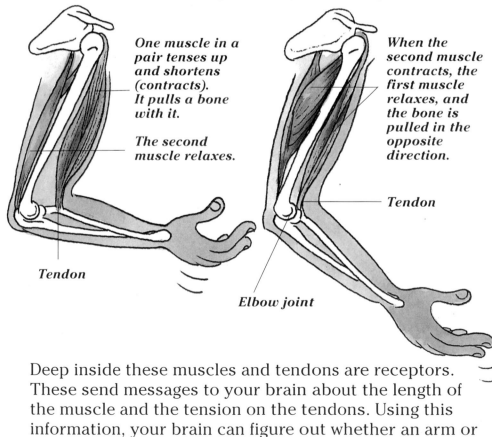

One muscle in a pair tenses up and shortens (contracts). It pulls a bone with it.

The second muscle relaxes.

When the second muscle contracts, the first muscle relaxes, and the bone is pulled in the opposite direction.

Tendon

Tendon

Elbow joint

Deep inside these muscles and tendons are receptors. These send messages to your brain about the length of the muscle and the tension on the tendons. Using this information, your brain can figure out whether an arm or leg is bent or straight.

MUSCLE MONITOR

Another way your brain knows the position of some parts of your body is by monitoring its own commands. It keeps track of instructions that have been sent to certain muscles.

You use this monitoring system for seeing. Your eye muscles move constantly. So the image on your retina is always jumping around. But you don't see it like this.

This is because your brain takes into account its commands to your eye muscles when interpreting the image on your retina.

For a link to a website where you can test your own body receptors with a couple of simple experiments, go to **www.usborne-quicklinks.com**

KEEP YOUR HEAD

Inside your inner ear you have a special system, called the vestibular system. It helps you keep your balance and tells you which direction you are moving in.

The vestibular system is made up of three tunnels, called semicircular canals, filled with liquid. At the end of each canal is a small swelling, covered with receptor cells, shaped like tiny hairs.

When you move, the liquid flows over the hairs and bends them. This bending is converted into electrical signals which are sent to the brain. Your brain figures out which way up you are, and where you are going, by analyzing these signals.

The vestibular system

Semicircular canals

Crista: small swelling covered with receptor hairs

In a car, your eyes tell your brain you are moving, but your vestibular system says you are still. This can make you car sick.

DIZZY SPELLS

Have you ever spun around in a circle so that when you stop you're so dizzy you fall over in a heap on the floor? This happens because the liquid in your semi-circular canals keeps on spinning after your body has stopped. The same thing happens with water in a glass.

Rotate a glass of water in your hand. The water in the glass will continue swirling after you have stopped moving the glass.

BRAIN BLINDNESS

There are several parts of your brain which deal with sight. If any of them is damaged, very strange things can happen to the way you see the outside world. Studying people with this kind of brain damage is very important to doctors. It can help them understand which part of the brain does what, when interpreting images on your retina.

FAMILIAR FACES?

People are better at recognizing faces than anything else. If you were shown 50 photos of faces one day, and then the same 50, plus 50 new ones the next day, you could tell the new from the old very easily.

But if you were to damage a tiny area on the right-hand side of your brain, you wouldn't be able to recognize any faces at all – not even your own.

The strange thing is that it is only faces that cause these difficulties. People with this problem can recognize someone by their clothes, voice or even handwriting. This makes scientists think there may be one part of the brain that is specially designed to recognize faces.

WHAT IS IT?

Brain damage to a similar area of the brain (on either the right or left side) can leave people being able to see objects without understanding what they are. Once they can touch something, they can identify it easily.

Someone brain damaged in this way looked at a mug and described it as a cylinder with a loop attached to one side. She didn't know what it was until she picked it up.

I think it's stopped!

An ex-doctor described a stethoscope as a long cord with a round thing at one end and two rigid cords attached at the other. The only thing he thought it could be was a watch.

54

LEFT BLINDNESS

 Damage to an area at the back of the right-hand side of the brain can stop people from 'seeing' anything on their left-hand side, even though their eyes are working perfectly. When asked to copy a drawing, these people will draw the right-hand side of the picture accurately, but leave out the left-hand side altogether.
They cannot even imagine a complete scene in their mind.

When asked to divide a horizontal line in half, a woman with left blindness drew a line near the right end of the line.

An Italian man was asked to pretend he was standing at one end of a famous square in Milan and to describe what he could see in front him. He described exactly the right-hand half of the square, but didn't mention the left.

BLIND SIGHT

 Each receptor in your retina sends messages to one particular point on your visual cortex (see page 36). If part of the visual cortex is damaged, the part of the retina which sends messages to that part of the cortex seems not to work. The person develops a big blind spot. If an object is positioned so it falls on this blind spot, that person will say they can't see it. The pictures below show an experiment carried out on hundreds of people with this brain damage. It proves that the objects ARE seen, but in a part of the brain the patients aren't aware of.

A shape is held up so its image falls on the patient's blind spot.

The patient claims not to be able to see it.

When asked to guess the shape, nearly all patients guess right.

ANIMAL SENSES

We think that we perceive the world as it really is. In fact, our senses give us a limited view of it. Many animals have sense organs that are very different from ours. This means that their experience of the world around them is different. They may be able to hear sounds too high for human ears, or smell scents that we wouldn't catch a whiff of.

FOLLOW THE SCENT

Many animals that hunt for food use their sense of smell to track down their prey. Sharks can smell even a tiny drop of blood in the ocean.

Ouch! Hope there are no sharks about!

Some dogs are one million times more sensitive to smells than we are. The police use specially trained dogs to help them hunt for missing people and to search for bombs, using their sense of smell. In 1925, a Doberman Pinscher tracked two thieves 160km (100 miles) across the desert in South Africa just by following their scent.

From smelling a piece of clothing, dogs can track down the person it belongs to.

SNIFF... SNIFF...

PROUD AS A PEACOCK

When a peacock wants to attract a mate, it shows off its bright feathers. Since this display would be wasted on a colorblind peahen, we can guess from this that they can see in color.

Me..? Vain..?

Most animals, however, don't have color vision. This is because they are descended from animals which hunted at night. Because colors can't be seen in the dark, color vision would have been useless.

SWAT THE FLY

If you've ever tried to swat a fly, you'll know how hard it is, even if you creep up on one from behind. There are two reasons for this. Flies, like all insects, receive visual information very quickly. This means they can react to things very fast. Flies also have large, curved eyes, giving them good all-around vision. So they can spot you coming from any direction.

But although they can avoid the swat, insects can't actually see very well. Their eyes are made up of lots of minute lenses, so they see the world as made up of hundreds of tiny dots.

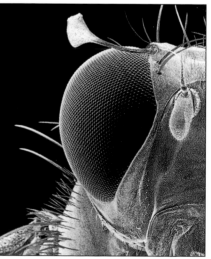

A fly's eye magnified more than 100 times

HEAT WAVE

A rattlesnake can track down prey by the heat it gives off. On either side of the snake's head is a pit containing heat receptors. These detect any increase in temperature in its surroundings, sensing the body heat of the snake's next victim.

BLIND AS A BAT

Bats have poor eyesight, but very good hearing. They can hear sounds too high for human ears to detect. Their sense of hearing helps them hunt for food and find their way in the dark.

Bats make high-pitched squeaks and then wait for echoes to bounce back off the objects in their path. From the echoes, they can tell the size and position of objects around them. This is called echo location.

Sound waves sent out

Echo returns

This bat can hear where its dinner is.

A SENSE OF DIRECTION

Birds that migrate to warmer places in winter have to travel thousands of miles. Scientists have discovered that they have a kind of internal "clock", which helps them do this. It is like an extra sense.

It allows them to use the Sun as a compass. Without it, the Sun would be no use as a guide, because it changes position throughout the day. But if you know it's six in the morning, you know the Sun is in the East, so you can use it to work out every other direction.

Migrating birds fly over thousands of miles. Many return to exactly the same spot year after year.

FASCINATING FACTS

A sperm whale can stun or even kill its prey by making loud noises.

A lizard called the tuatura has a third eye on top of its brain.

Blue whales have eyes as big as footballs.

Grasshoppers' ears are in their knees.

The African elephant has the biggest nose of any mammal. A large male's trunk is about 2.5m (8ft) from base to tip.

For a link to a website where you can find out lots more information about animal senses and hear a whale's song, go to **www.usborne-quicklinks.com**

LEARNING TO SENSE

When a baby is born, most of its senses organs are fully formed. But babies can't sense things in the way older children and adults can. This is because a baby's brain has not yet developed enough to analyze the messages it receives from its senses.

Scientists think that babies' senses are programmed to develop in a certain way, but that other things can affect them. How and where a baby is brought up, and what it is taught, can make a difference to how its senses work.

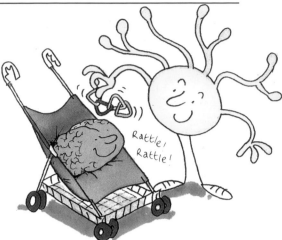

BLURRY FACES

Newborn babies' eyes still have some growing to do before they are perfectly formed. So they can't see as well as you.

This is what a face looks like to a newborn baby.

By 6 months, it can see as well as an adult.

TOWN AND COUNTRY

Comparing children who live in town and the country, it is possible to see differences in how their senses work because of where they live. For example, a city child may judge how fast a car is traveling, through sight, sound and touch (the feel of wind as a car whizzes past). This will help him cross a busy road. A child from a farm may not be able do this so easily. But she may be able to tell a sheep apart from its flock.

For a link to a website where you can listen to an unusual version of a well-known story, go to **www.usborne-quicklinks.com**

LEARNING TO HEAR

As they get older, babies' brains learn only to tune into sounds which are familiar. This means they lose the ability to say and hear sounds they don't hear very often. This is why, for example, Japanese children can't hear or say the difference between l and r sounds. In Japanese there is only one sound for these, somewhere in between the two.

At 6 months, all babies, from all over the world, babble the same sounds.

But by 12 months or so, babies only babble the sounds they hear in their own language.

DOUBLE DUTCH

When foreigners talk, it can sound as if they are talking really quickly. It's even hard to tell where one word ends and another begins. They are probably speaking no faster than you do, but your brain has not learned to 'hear' their language.

PERFECT SENSE

If you use a particular sense a lot for a particular job, your brain gets better at interpreting the information it receives.

Expert wine-tasters can tell if a wine is from Italy, U.S.A., Australia, or a particular region of France, which variety of grapes were used to make it, and even what year it was made.

Perfume companies employ people with a trained sense of smell to test out their perfumes for them.

Bird watchers can identify birds which, to the untrained eye, just look like brown blobs.

JUST AN ILLUSION?

In general your senses do a pretty good job of telling you about the world around you. But not always. Sometimes they can give you false information. These are called illusions. Most illusions occur because of the way your senses work. This means scientists can get useful information about your senses by studying illusions.

ELECTRICAL FAULTS

Your receptors fire off more electrical pulses when they are first activated. But once they are used to a situation, they fire off fewer. Many illusions, such as the one below, happen because of this.

Stare at this flower for 30 seconds.

Now look at a sheet of white paper, and blink a few times.

The image of a flower should appear, but with red petals.

This happens because white has red and green light in it. When you look at the white paper, your red receptors are responding for the first time, so they fire off lots of pulses. Your green receptors have already been responding, so they fire off fewer. This makes you see red petals.

SOUND ILLUSION

When a police car speeds past you, the pitch of the siren suddenly changes from high to low, although the noise of the siren is actually quite regular. This happens because the closer together sound waves are, the higher the pitch of the sound.

The sound waves coming from the police car bunch up together as the car moves toward you, so the siren sounds higher than it really is. As the car moves away, the waves behind it appear to spread out. This makes the sound seem lower.

TOUCH ILLUSION

Your sense of touch is fooled in this experiment because you are using your fingers in a way your brain is not used to.

1. Take a small round lid or a large coin, more than 2.5cm (1in) across.

2. Hold it between your left thumb and forefinger.

3. Close your eyes and turn the lid with the right thumb and forefinger.

You should get the feeling that a perfectly round object is oval-shaped.

This illusion works better for some people than others. The larger the object and the faster you turn it, the stronger the illusion should be.

PERSPECTIVE ILLUSIONS

Your brain turns the flat images on your retina into 3-dimensional scenes. Many visual illusions happen when your brain tries to do the same thing to flat pictures on a page.

Try these quiz questions and see how your brain can trick you. Don't use a ruler to help you.

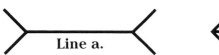 **Line a.**

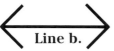 **Line b.**

1. Which line is longer, a or b?
(Answer on page 98.)

2. Which man in this picture is the tallest?
(Answer on page 98.)

WHERE'S THE SQUARE?

Sometimes when you look at a pattern, regular shapes can appear distorted. Because the combination of shapes is unfamiliar, your brain doesn't know how to interpret the image.

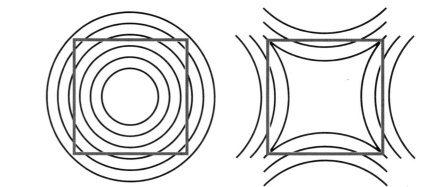

3. Which of the patterns contains a perfect square? (Answer on page 98.)

DUCK OR RABBIT?

Sometimes, something you look at can be one of two things. But there are not strong enough features for your brain to make its mind up which one is correct.

This picture can be either an old or a young woman.

This simple line picture can be either a duck or a rabbit.

This is a vase or two faces in profile. Which did you see first?

SMART MACHINES

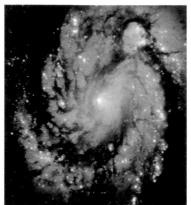

For hundreds of years, people have used equipment to help them see and hear things which they couldn't using only their own senses. Today, computers have been developed that not only improve our senses, but can also actually sense things for themselves.

The Hubble telescope allows human eyes to see millions of light years away. This is a spiral galaxy, about 50 million light years from Earth.

COMPUTER SENSES

Computers can sense things just as you can. But, instead of sense organs, they have input hardware. Input hardware converts information into a series of electrical signals, called binary code. Binary code can be analyzed by a computer's processor. This is like the brain of the computer.

A computer processor

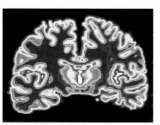

A brain cross-section

This picture shows how computers work in a similar way to your senses and brain.

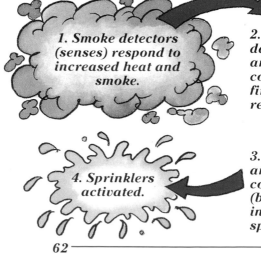

1. Smoke detectors (senses) respond to increased heat and smoke.

2. Detectors convert data about smoke and heat into binary code (like pulses fired off by receptors).

3. Binary code analyzed by computer processor (brain) and instructions sent to sprinkler system.

4. Sprinklers activated.

COMPUTER SIGHT

Scanners are the "eyes" of a computer. They can see images or printed text, and translate them into binary code. The computer reads the code and can recreate the image or text on screen.

X-ray scanners are used in hospitals to see inside the human body. If enough x-rays are taken, a computer can build up 3-D images of the inside of your body.

Ah, looks like 0110100110!

Hand-held scanner

Although scanners can see things you can't, scientists have yet to make a computer that works as well as the brain. For example, computers are pretty bad at recognizing faces - a task which humans can perform at only a few months old.

TOUCH SENSITIVE

A touchscreen is a computer screen with a built-in system of wires which can feel touch. Touchscreens are often used in information centers. You can select options and give commands by pressing different areas on a screen.

VOICE RECOGNITION

Some computers can hear. They convert sound waves into electrical signals. But computers have as many problems with voices as they do with faces. No one has yet developed a computer which can easily follow instructions from people's voices. This is because people speak in so many different ways it would need a very complicated processor.

VIRTUAL REALITY SENSES

Computers can create worlds that don't exist, and make you believe you can see, hear and touch them. This is called virtual reality. In the virtual world, just by putting on some special equipment, you could land a plane (pilots train like this) or creep through a haunted house.

The girl in the picture is wearing a headset and data gloves. Sensors in the headset pick up head movements. They convert this information into signals which are sent to a computer. As her head moves, the computer changes the image on the headset's screens, and adjusts the sound. Sensors in her data gloves detect hand movements in the same way.

This girl is wearing special virtual reality gear.

The headset has two screens in front of each eye. 3-D scenes are projected onto these screens.

Data gloves

The headset also contains earphones with stereo sound.

A GLOSSARY

Here's a reminder of what some of the difficult words in this book mean.

Amplitude – The loudness of a sound.

Binocular vision – The result of your eyes being at slightly different positions. They produce two images that are joined together by your brain to help you see in 3-D.

Color blindness – An eye defect that makes people unable to tell the difference between certain colors.

Cones – Cells in your eye that respond to different colors.

Constancy mechanisms – The methods that your brain uses to recognize the real size, color or brightness of the objects you look at.

Cortex – The area of your brain where information that comes from your senses is processed.

Decibels – The unit in which loudness is measured.

Endorphins – Chemicals released by your brain to control pain.

Harmonics – The mixture of high and low, loud and quiet notes that make up a sound.

Pitch – How high or low a sound is.

Receptors – The parts of your sense organs that respond to what happens in the world around you.

Referred pain – Pain felt in one part of your body, when an organ in another part has been damaged.

Rods – Cells in your eye that respond to bright or dim light.

Sense organs – These are the parts of your body which can sense things. Your sense organs include your tongue, eyes, nose, ears and skin.

Thalamus – A small area of your brain which receives information from your senses. The thalamus then sends this information to the particular part of your cortex that deals with it.

Vestibular system – A system of tunnels in your inner ear that controls your balance and your sense of direction.

DID YOU KNOW?

Dogs have a very good sense of smell. One breed, called a German Shepherd, has 44 times more smell receptors in its nose than a human being.

Animals that feed at night use their senses to find their way in the dark. Some, such as moles and bats, have poor eyesight, but very good hearing and smell. Others, such as bush babies and tarsiers, have huge eyes.

An Eastern tarsier (a type of monkey) has eyes that are 17mm (0.6in) across. If a human's eyes were the same size in proportion to its body, they would be as big as grapefruits.

Cats have a special layer at the back of their eyes which reflects light. This means they can make much better use of light than human beings. They can hunt well at night because they are able to see in very dim light.

Understanding your
Muscles
& Bones

INTRODUCTION

CONTENTS

66 Introduction

68 Your skeleton

70 Skeletal muscles

72 Joints

74 Moving parts

76 Looking inside

78 Your heart

80 Body batteries

82 Strength and sports

84 Involuntary muscles

86 Muscle talk

88 Growing

90 Broken bones

92 Muscle trouble

94 Prehistoric bones

95 Amazing facts

96 Index

This part of the book is all about muscles and bones. They work together as a team allowing you to run, skip, play sports or just turn on the TV. Without them you would be a heap of body matter, doomed to live out your life as a motionless blob.

Look at your hands. Think of all the different things you can do with them, from gripping tightly to playing the violin.

It is an intricate system of muscles and bones which gives you this flexibility.

This shows the muscles in your hand.

Food for Thought

If you eat meat you are actually eating animal muscle.

BRAIN POWER

Your muscles can only move under instructions from your brain. Your ability and skill at controlling your movement is called coordination.

Tiny babies' movements are very limited as they have no coordination.

A toddler's brain is just beginning to develop coordination.

Soccer players have to coordinate many parts of their bodies to hit a ball on target.

CELLS AND TISSUES

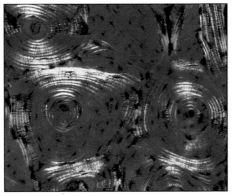

Bone tissue

Muscles and bones are types of body tissues. Tissue is made up of cells, the building blocks of your body. Bone tissue is made up of bone cells. Muscle tissue is made up of muscle cells. There are several types of muscle tissue. You will find out about them all in this part of the book.

EXPERTS

Some doctors and scientists are experts in muscles and bones.

Osteologists study the structure and function of bones.

Physiotherapists try to rebuild your muscles if they have been weakened by disease or an accident.

Osteopaths try to heal many illnesses by massaging and moving muscles and bones.

Finger Facts

There are 27 bones and 37 muscles in your hand.

Each finger contains three bones and each thumb two. This allows you to curve them around a pen to write.

Did you hear the one about the skeleton who wouldn't do his homework? He was bone idle!

YOUR SKELETON

Your skeleton is a frame which gives shape to all the soft jelly-like parts of your body and stops them from falling in a flaccid heap on the floor. It also forms a protective cage around your organs, such as your heart, and prevents them from getting damaged by knocks and bumps.

You might expect adults to have more bones than babies, but actually it is the other way around. There are 206 bones in a fully-grown human skeleton, but more than 300 in a baby's. This is because, as we grow, some of our bones fuse together to form bigger bones.

Nearly half of all your bones are in your hands and feet.

Distal phalanx
Middle phalanx
Proximal phalanx
Metacarpal

Carpus

Bone from your ear

The smallest bone in your body is only 3mm (0.12in) long. It is found in your ear and it vibrates to let you hear.

An adult skeleton

Skull

Your skull is like a crash helmet protecting your brain.

Humerus

NECKTIES

Your neck contains seven vertebrae. How many do you think a giraffe has? (Answer on page 98.)

Clavicle (collar bone)

Ulna
Radius

Sternum (breast bone)

Scapula (shoulder blade)

12 ribs form a cage to protect your lungs and heart.

Vertebral column (spine or backbone) made up of 33 vertebrae

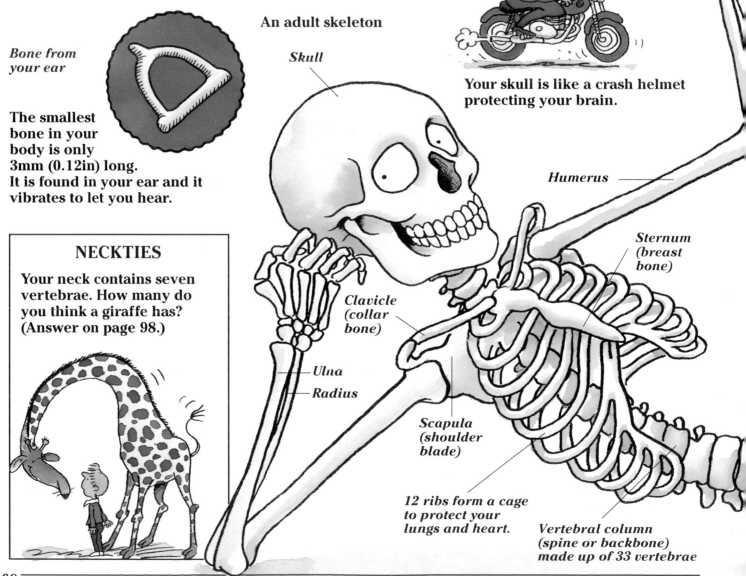

SPINES, SHELLS AND BODY BAGS

Humans are vertebrates. This means we have a backbone, or spine. Your spine is your body's main support. Some other vertebrates share the same type of skeleton as humans. So although you look totally different from a pig, cat or rabbit, your skeleton follows the same general pattern as theirs, with a skull, four limbs and a spine.

Bird skeleton

Animals without spines are called invertebrates. Some invertebrates, such as crabs, have an exoskeleton - a hard shell-like skeleton outside their bodies.

Crab

Fish and birds are vertebrates too, but their skeletons follow a different pattern.

Cat skeleton

Other, simpler animals, such as worms, have no skeleton. All their body parts are contained in a bag.

Earthworm

Femur - the biggest bone in the body

Ilium

Sacrum   *Ischium*

Pubis

Patella (knee cap)

Tibia

Fibula

Middle phalanx

Distal phalanx

Proximal phalanx

Metatarsal

Coccyx. Some scientists think your coccyx is all that remains of a tail which disappeared as humans evolved.

Tarsus

SKELETAL MUSCLES

Each person has about 600 muscles, called skeletal muscles, attached to their skeleton. They keep the skeleton upright and give your bones the power to move. Skeletal muscles are also called voluntary muscles. This is because you can consciously control their movement. Skeletal muscles have Latin names, which are used and understood by scientists all over the world.

MUSCLE FUEL

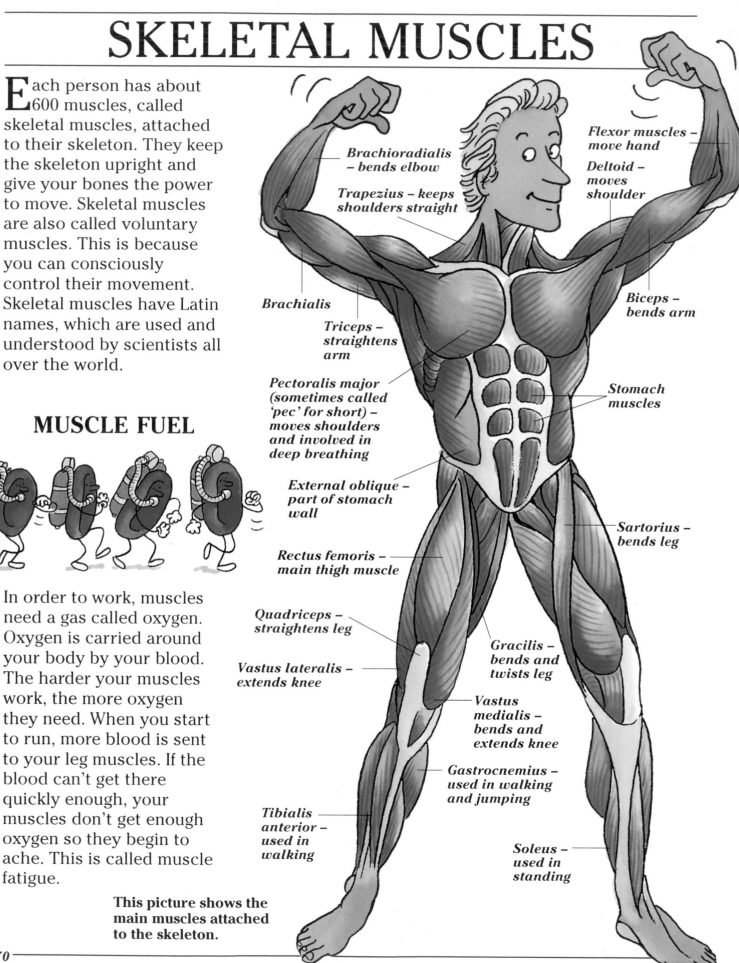

In order to work, muscles need a gas called oxygen. Oxygen is carried around your body by your blood. The harder your muscles work, the more oxygen they need. When you start to run, more blood is sent to your leg muscles. If the blood can't get there quickly enough, your muscles don't get enough oxygen so they begin to ache. This is called muscle fatigue.

This picture shows the main muscles attached to the skeleton.

Brachioradialis – bends elbow

Trapezius – keeps shoulders straight

Flexor muscles – move hand

Deltoid – moves shoulder

Brachialis

Biceps – bends arm

Triceps – straightens arm

Pectoralis major (sometimes called 'pec' for short) – moves shoulders and involved in deep breathing

Stomach muscles

External oblique – part of stomach wall

Rectus femoris – main thigh muscle

Sartorius – bends leg

Quadriceps – straightens leg

Gracilis – bends and twists leg

Vastus lateralis – extends knee

Vastus medialis – bends and extends knee

Gastrocnemius – used in walking and jumping

Tibialis anterior – used in walking

Soleus – used in standing

MAKING FACES

You have over 40 muscles in your face - more than any other animal. This makes you very expressive. You can communicate many emotions through your facial expressions. Can you tell what these people are feeling just by looking at their faces?

CLENCH TEST

Hold your hand above your head, and see how many times you can clench and unclench your fist before it begins to ache. Once the aching has stopped, try the same thing with your other hand down by your side.

You should be able to manage more clenches with your hand hanging down because it is easier for blood to flow down into your hand rather than upward.

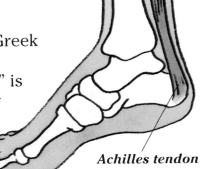

TALES OF CLASSIC TENDONS

Muscles are attached to bones by tough, inelastic (non-stretchy) bands called tendons. The tendon which attaches the muscle to the back of your foot is easily damaged during exercise. It is called the Achilles tendon, after an ancient Greek story (see below). The expression "Achilles heel" is also used to describe any weakness in an otherwise strong person.

Achilles tendon

1. Thetis dipped her baby son Achilles in a sacred river to protect him from danger.

2. The only spot which stayed dry was his heel. This became his weak spot.

3. Achilles was eventually killed by an arrow wound in his heel.

JOINTS

Joints are the places where your bones meet. Try moving your elbow and knee joints. They work like hinges, opening and closing. Now move your shoulder and hip. They open and close like your elbow or knee but also move around in a circle. Different joints allow you to move in different ways, because the bones are joined differently.

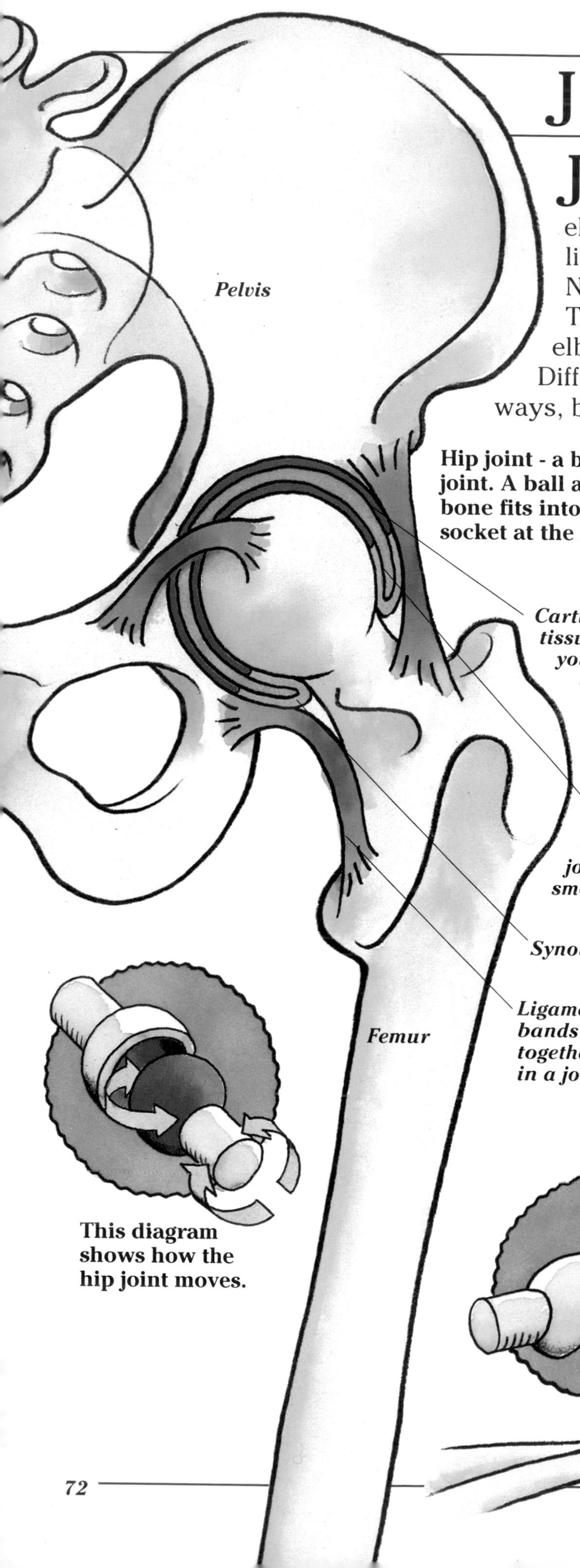

Pelvis

Hip joint - a ball and socket joint. A ball at the end of one bone fits into a cup-like socket at the end of another.

Knee joint - a hinge joint

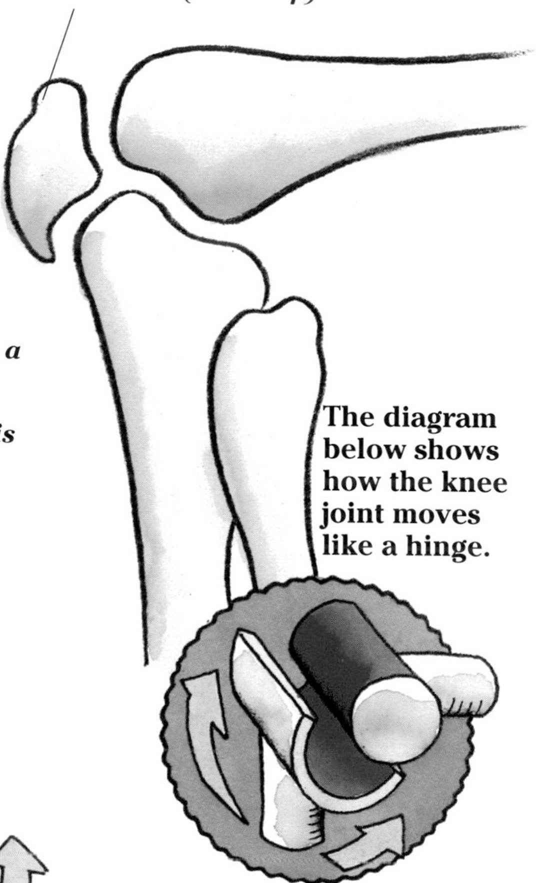

Patella (knee cap)

Cartilage is a soft tissue at the end of your bones which cushions your joints and acts as a shock absorber.

Synovial fluid is a special liquid which "oils" the joint so movement is smooth.

Synovial membrane

Ligaments are tough bands which join together the bones in a joint.

Femur

The diagram below shows how the knee joint moves like a hinge.

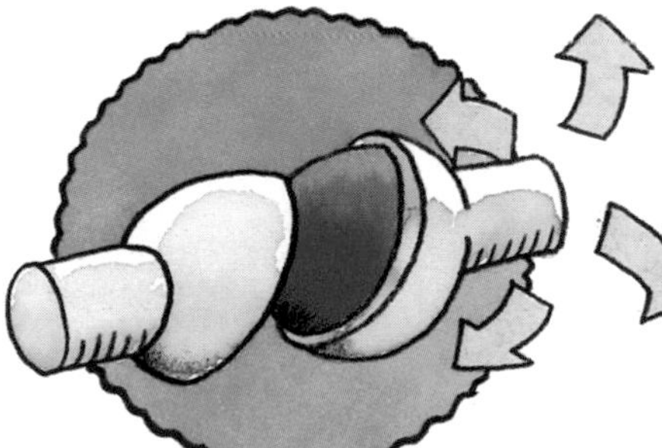

This diagram shows how the hip joint moves.

Wrist joint - a sliding joint. The bones slide over each other (see left).

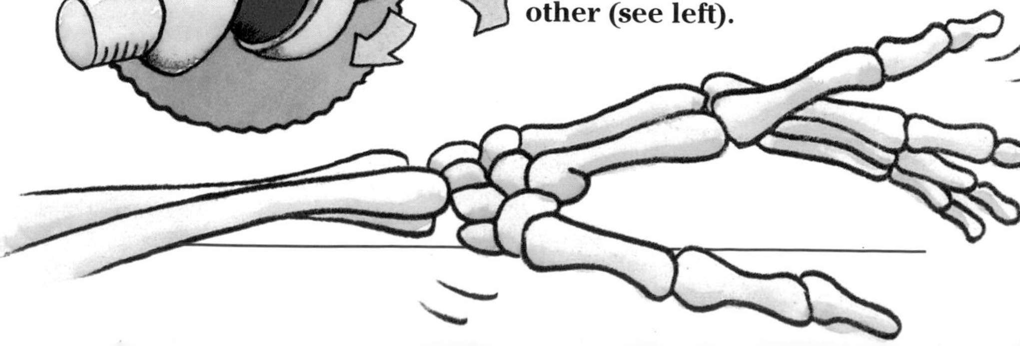

CREAKY JOINTS

Joints are one of your body's trouble spots. They are under almost constant pressure and tend, especially as you grow older, to develop faults.

It is easy to sprain a wrist or ankle. Sprains happen if you twist a joint suddenly and tear or snap the tendon or ligament. Sprains can be painful, but they usually heal themselves given time and rest.

Arthritis is a disease which causes joints to swell up so they become difficult to move. There are two kinds of arthritis.

In osteoarthritis, the cartilage in the joints wears down. So instead of moving smoothly, the bones grind against each other.

In rheumatoid arthritis, the bones in the joint become fused together. This makes any movement impossible.

Osteoarthritis usually happens in older people, but people of any age can get rheumatoid arthritis.

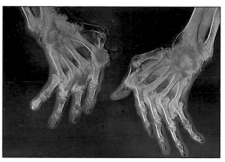

Finger joints crippled by rheumatoid arthritis

Joints are not designed to withstand the vigorous training that some sports demand. When a joint is used too much, its lining can get inflamed and it can produce too much synovial fluid (see opposite).

People who use their arms a lot, such as tennis players, can suffer from "tennis elbow" - an inflamed elbow caused by overuse.

SPLITS AND BACK FLIPS

Gymnasts and acrobats can bend their bodies into shapes which could put any normal person in the hospital! This is because they are very supple. This means they have more movement in their joints. You can become supple by doing special exercises which stretch your body beyond its normal range. These must only be done with a trained instructor, or you could hurt yourself badly.

MOVING PARTS

Every time you move, your brain, nerves, muscles and bones all work together in a highly complex way.

Your brain receives information from your senses. This is evaluated and a decision is made about whether the body should move in response. If movement is needed, messages, called impulses, are sent down nerve cells, usually via your spinal cord, to nerve endings at your muscles. The nerve endings stimulate the muscles into action and you move.

1. Information from the senses (in this case the eyes) is sent to the brain.

2. The brain decides what response is needed.

3. The brain sends out impulses to muscles.

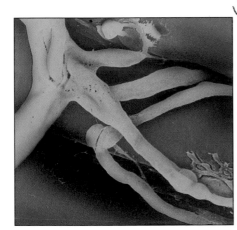

This photograph shows a magnified image of nerve endings (in yellow) meeting a muscle (in pink).

4. The muscles are stimulated and the body moves in response.

PAIRS

Muscles move bones by contracting. When a muscle contracts, it tightens and shortens, pulling a bone with it. But muscles can't push. This means they have to work in pairs: one to pull a bone one way, and the other to pull it back again.

Biceps contracts - forearm moves up.

Triceps relaxes.

Biceps relaxes.

Triceps contracts - forearm moves down.

These pictures show how your biceps and triceps in your upper arm work together to move your forearm.

For a link to a website where you can play a game to see how your muscles fit together, go to www.usborne-quicklinks.com

LEONARDO

In Europe, between five and six hundred years ago, people became very interested in the human body and how it worked. Artists, such as Leonardo da Vinci, studied the body and muscles in particular. Leonardo filled many sketch books with his drawings of the structure of muscles.

RUMBLING THUMBS

Put both thumbs gently into your ears and then clench your fists. As you clench you should hear a rumbling sound. This is the sound of your muscles vibrating as they contract. The tighter you clench, the more your muscles contract and the louder the rumbling sounds will become.

REFLEX ACTIONS

Actions such as dancing or pressing a doorbell are voluntary. This means they are the result of a conscious decision. But there are some actions which are involuntary, over which you have no control. One special kind of involuntary action is a reflex action.

A reflex action is your body's response to a dangerous situation. It happens so quickly you are not aware of what your body has done until the action is over. This is because the message to act comes from the spinal cord and not your brain. The brain is only told what has gone on after it has happened.

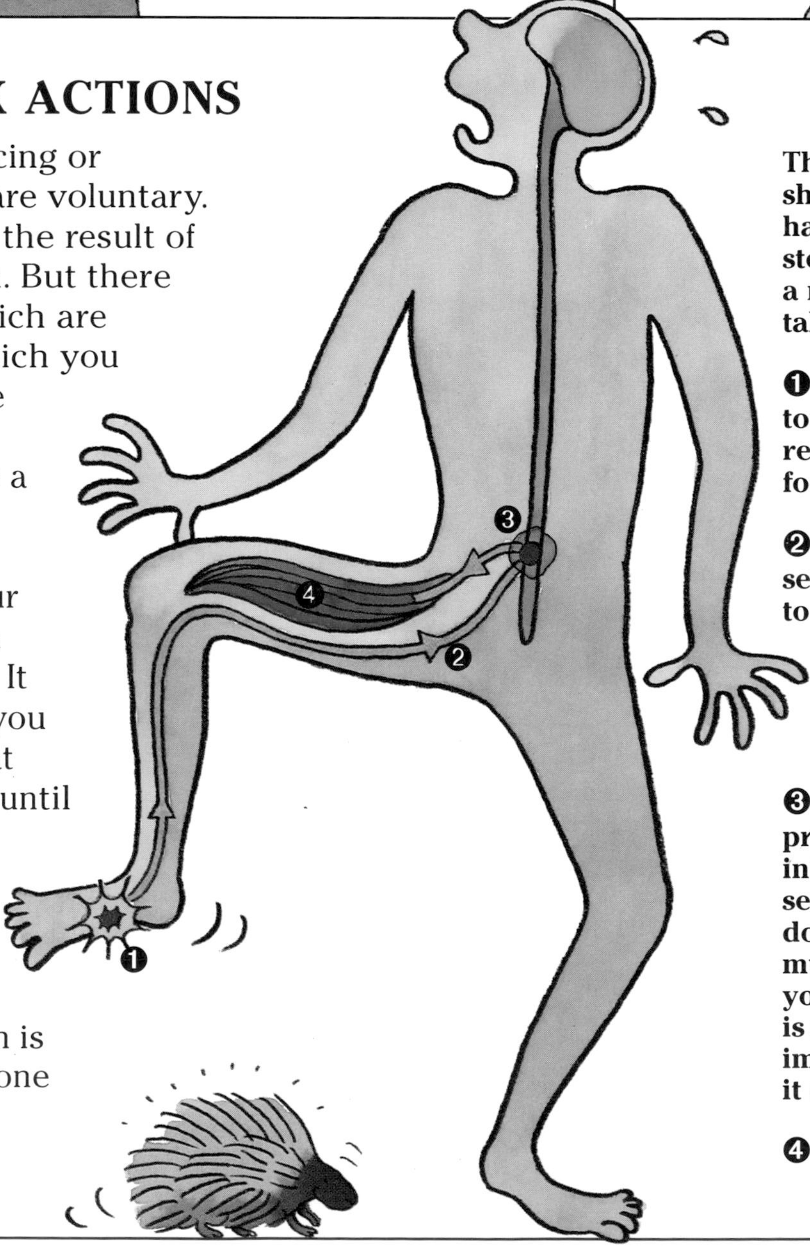

This diagram shows you what happens when you step on a spike and a reflex action takes place.

❶ Sharp spike touches pain receptors in your foot.

❷ Nerve endings send information to special nerve cells in the spinal cord. The brain is bypassed.

❸ The spinal cord processes this information and sends impulses down to the muscle that moves your leg. The brain is also sent impulses informing it of the action.

❹ Leg moves.

75

LOOKING INSIDE

Bone might look dead, but it is in fact very much alive. Bone is a living tissue made up of bone cells, blood vessels and nerves. Bone can grow, repair itself and hurt if it is damaged.

WHAT MAKES BONES STRONG?

Compact bone is the hard, strong part of your bones. It is made strong by a mineral called calcium. We get calcium from the food we eat. It is very important for babies to drink a lot of milk, because it contains the calcium that will make their bones harden. Calcium also makes your teeth hard.

Inside a bone

The outer part of the bone is called compact bone. This is the hard part.

In the middle of the bone is a tube containing a jelly called marrow.

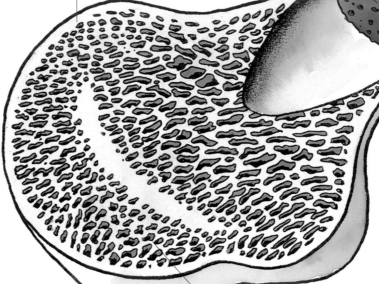

At the end of bones, beneath the compact bone, is a lighter substance called spongy bone. It isn't spongy to touch, but it does look a little like sponge.

BENDING BONES

Take a chicken leg bone and put it in a bowl filled with vinegar. Leave it for two to three days.

Pour away the vinegar, wash the bone with water and then try to bend it.

The bone bends because the vinegar has dissolved the calcium in it.

BLOOD FACTORY

In the marrow of a few bones, over two million new blood cells are being made every second. This is a vital task, essential for your survival.

You have two types of blood cells: red and white. Red blood cells carry essential substances around your body. White blood cells are germ-busters. They defend against disease. There are two main types of white blood cells: phagocytes and lymphocytes. Phagocytes kill germ cells by eating them. Lymphocytes kill germs by blasting them with chemicals called antibodies.

Yummy!

Phagocyte

FRRZZZT!

TAKE THAT GERMS!

FRRZZZT!

Lymphocyte

OLD BONES

This human skeleton is over 1500 years old. The rest of the body would have decomposed soon after the person died.

The bones in the skeleton above are not alive like yours. The bone cells have died and all that is left is calcium and other minerals. They would crumble if handled roughly.

This was obviously a left handed vegetarian who enjoyed football and...

Archeologists can find out a lot about how people lived by looking at old bones. By testing the minerals in skeletons, they can figure out what people ate. From this they may be able to suggest whether they were hunters or farmers, and so what tools and weapons they made.

MUSCLE TISSUE

Muscles are made of muscle tissue. There are three different types of muscle tissue: heart muscle, smooth muscle and striped muscle. Heart muscle makes up your heart. Smooth muscle makes up the muscles in the walls of your gut and other organs. Striped muscle makes up your skeletal muscles.

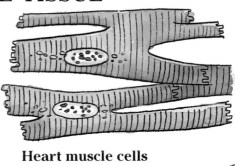

Heart muscle cells

Smooth muscle cell

Striped muscle consists of bundles of cords. Each cord is made up of two kinds of strands, or filaments. Muscles contract when these filaments overlap and lock into each other.

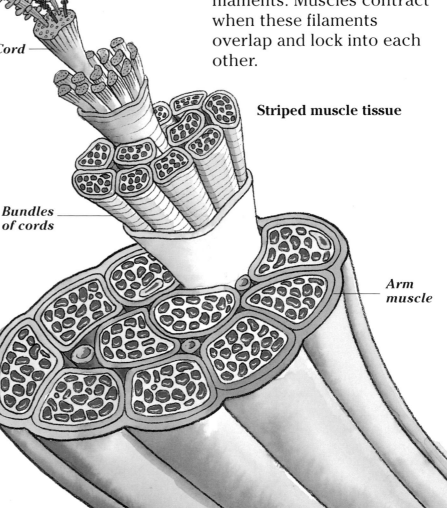

Filament

Cord

Striped muscle tissue

Bundles of cords

Arm muscle

YOUR HEART

Your heart is an amazingly efficient muscle. It never stops working and it never gets tired. It is the life force of your body, pumping blood to all corners of your body.

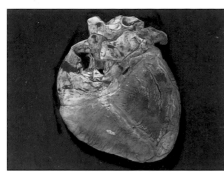

Photograph of a human heart

HEART BEATS

If you put your hand on your chest you can feel your heart beating. Each beat is caused by your heart muscle contracting (tightening) and then relaxing. This happens about 100,000 times a day or about 70 times a minute.

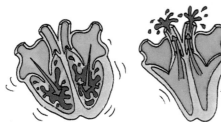

Heart relaxes   Heart contracts

When your heart is relaxed, blood flows into it from the veins. When it contracts, blood is pumped out of the heart into the arteries. If your body is working very hard it needs more blood, so your heart beats faster.

Your heart is divided into two halves: right and left. Each half is made up of an upper "chamber" called an atrium and a lower "chamber" called a ventricle.

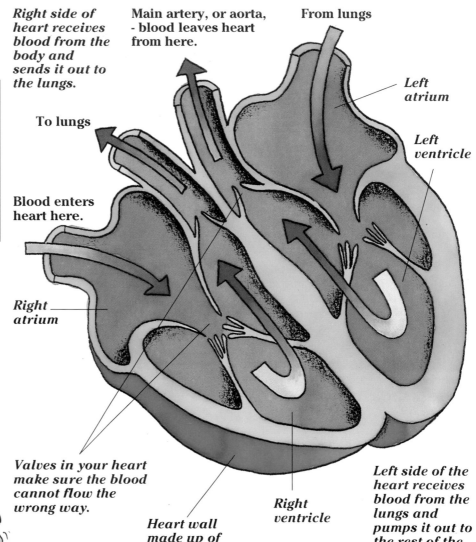

Right side of heart receives blood from the body and sends it out to the lungs.

Main artery, or aorta, - blood leaves heart from here.

From lungs

To lungs

Blood enters heart here.

Left atrium

Left ventricle

Right atrium

Valves in your heart make sure the blood cannot flow the wrong way.

Heart wall made up of heart muscle

Right ventricle

Left side of the heart receives blood from the lungs and pumps it out to the rest of the body.

LOVEHEARTS

Many people think that their emotions come from their heart, but this is nonsense. It is your brain that allows you to fall in love or get upset. Feeling something strongly does however often make your heart beat faster. This is because when you get excited your body works harder and so your heart needs to pump more blood.

For a link to a website where you can see a picture of blood flowing through the heart, go to **www**.usborne-quicklinks.com

BLOOD FLOW

Your body contains about a small bucketful of blood. Blood is used to transport things around your body. It circulates endlessly, delivering oxygen, food and chemicals wherever they are needed. It also picks up poisonous waste and carries it to where it can be dealt with safely.

Your blood is transported in thin tubes called blood vessels. There are three types of blood vessels: arteries, veins and capillaries. Arteries carry blood away from the heart, veins carry blood to the heart, and capillaries join arteries to veins.

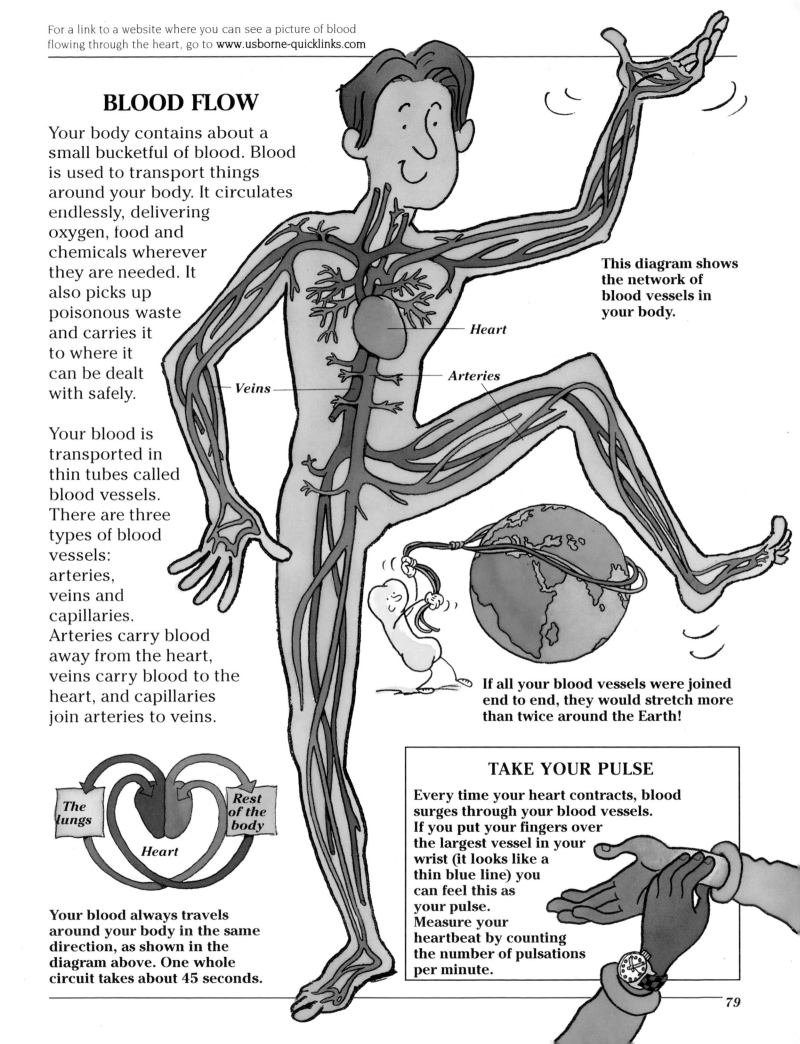

This diagram shows the network of blood vessels in your body.

Heart

Arteries

Veins

If all your blood vessels were joined end to end, they would stretch more than twice around the Earth!

The lungs

Heart

Rest of the body

Your blood always travels around your body in the same direction, as shown in the diagram above. One whole circuit takes about 45 seconds.

TAKE YOUR PULSE

Every time your heart contracts, blood surges through your blood vessels. If you put your fingers over the largest vessel in your wrist (it looks like a thin blue line) you can feel this as your pulse. Measure your heartbeat by counting the number of pulsations per minute.

BODY BATTERIES

Inside all your muscles is a vital substance called ATP. ATP is like a battery that stores energy. Your muscles can only contract if they have a constant supply of ATP. This is provided by a process called respiration. The more energetic you are, the more ATP you need. Without it your muscles are useless.

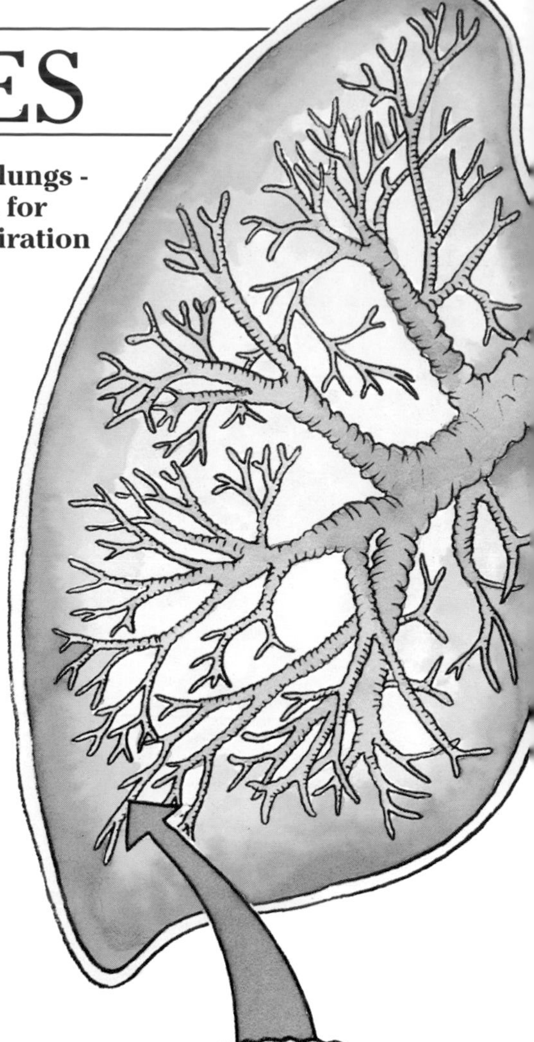

The lungs - vital for respiration

ATP AND OXYGEN

During respiration, food, which your body converts into a sort of sugar, is broken down in your muscle cells.

There are two types of respiration: aerobic and anaerobic. Aerobic means 'with oxygen'. During aerobic respiration, food reacts with oxygen to produce ATP. Your muscles get the oxygen they need for aerobic respiration from the air you breathe into your lungs.

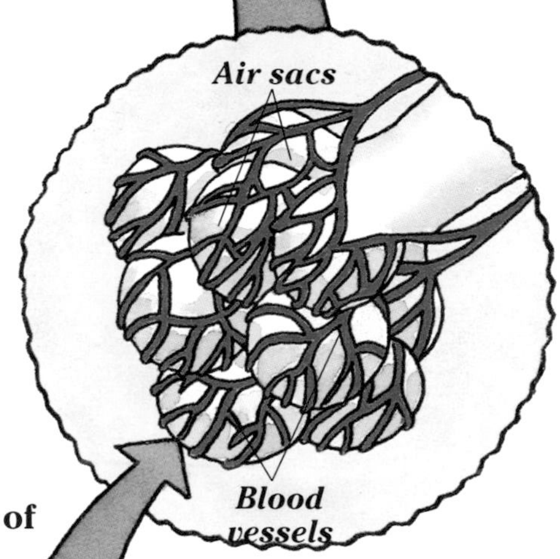

When you breathe in, millions of air sacs in your lungs fill with air. Each of these air sacs is surrounded by masses of tiny blood vessels.

Air sacs

Blood vessels

Oxygen passes through the walls of the sacs into your blood. Carbon dioxide passes from the blood back into the lungs. It is then breathed out.

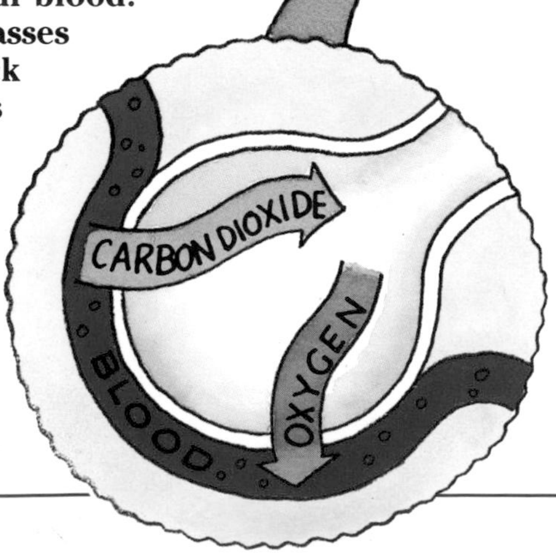

In aerobic respiration, oxygen and sugar react together. After some popping and fizzing, ATP, water and carbon dioxide are produced.

Once the oxygen is in your blood, it is pumped to your muscles by your heart.

CARBON DIOXIDE

BLOOD

OXYGEN

80

READY, STEADY, GO

Your heart can easily pump enough oxygen to your muscles to provide the energy you need for the less strenuous activities of everyday life, such as climbing upstairs. If you start dancing or skipping, your heart should still be able to supply enough oxygen for aerobic respiration. By beating faster it can get more oxygen to your muscles more quickly. But if you exercise really intensely, your heart won't be able to keep up. Then the anaerobic system takes over and ATP is produced without oxygen. The problem with anaerobic respiration is that it also produces a poison, called lactic acid. If lactic acid builds up in your muscles, it can give you muscle fatigue and stop you from moving altogether.

Anaerobic respiration is most useful when you need a lot of energy very quickly for a short amount of time: for example, for sprinting.

If you were running a mini-marathon, you would also need a lot of energy, but supplied more steadily, over a long period of time. For this you couldn't rely on anaerobic respiration, because the lactic acid would soon paralyze your muscles. Because of this, long-distance runners do special aerobic training to make their aerobic system more effective.

Step, two, three ... feel the burn, two, three ...

Aerobics is a popular form of aerobic training.

Aerobic training increases the oxygen in the blood and makes the heart beat stronger. Aerobic training must be hard, but not exhausting, and must be kept up regularly.

ENERGETIC ANIMALS

During respiration, some energy is released as heat. This is why you get hot when you exercise.

Other animals produce additional forms of energy. Some eels, for example, produce electricity. Glowworms produce light.

STRENGTH AND SPORTS

The more you use your muscles, the bigger they grow. And the bigger they grow, the stronger you will become. Strength is a very important part of many sports, and training to become stronger is vital in the preparation of any athlete.

TYPES OF STRENGTH

Strength is the ability of your muscles to exert a force on something to make it move. Strength always involves overcoming another force which doesn't want to move (either your own body or another object). This force is called resistance.

Force

Resistance

Different activities in sports require different types of strength.

General strength is the strength of all your muscles and how well they work together as a system. All sportsmen and women must have a good level of general strength.

Maximum strength is the greatest force which one muscle can exert.

Specific strength is the ability of particular muscles to perform a specific type of movement.

These athletes have a lot of specific strength in their arms.

Explosive strength is the ability of muscles to contract very quickly.

You use explosive strength to jump, throw and sprint.

Long distance running requires a lot of endurance.

Strength endurance is the ability to exert a force over a long period of time.

QUIZ

Which types of strength are most important in these activities? Answers on page 98.

1. Weight lifting

2. Competition to see who can do the most push-ups.

3. 100m sprint

BASIC TRAINING

Any training to make you stronger must follow three basic rules.

1. Muscles must be made to contract against a greater resistance than they are used to.

2. Resistance must be increased as training continues.

3. If training for a particular sport, the correct muscles must be exercised. If you want to be a runner, there is no point in doing hundreds of arm exercises.

Increase in muscle strength isn't permanent. If training is stopped, Mr. Powerhouse will eventually become Mr. Puny. The good news is that it takes about three times as long to lose new muscle strength as it does to gain it.

STRENGTH FROM STRESS

1. Exercise

2. Stress

3. Recovery

4. Stronger muscles

How does exercise make your muscles bigger? The answer is that it doesn't. It is the process of recovery from exercise that actually makes your muscles grow.

When you are training, two things are happening. First, your muscles are contracting more. Second, more chemical reactions are taking place in the muscle cells (see page 80). These two things put the body under stress, and it is the steps that are taken to stop this stress that make the muscle stronger.

Muscles don't actually start to get bigger until 24 to 48 hours after training has ended. Then individual muscle filaments begin to grow in thickness, so they can contract more strongly.

GREEK GOLDS

The Ancient Greek athlete, Milo of Cortona, used to lift a particular calf above his head daily. As the calf grew older and heavier, the resistance increased and

Milo's muscles grew stronger. Milo went on to win the top prize at six Olympic Games. This is one of the earliest examples of really effective training.

INVOLUNTARY MUSCLES

If you had to take care of the working of all your muscles all the time, you wouldn't have time to do anything else. Luckily, a part of your brain can control a whole host of muscles, without you being aware of it. These muscles are called involuntary muscles. You can take over control of some involuntary muscles when you need to.

DIGESTIVE MUSCLES

Your digestive system turns the food you eat into substances that can be absorbed into your blood. Involuntary muscles play a vital role in this process.

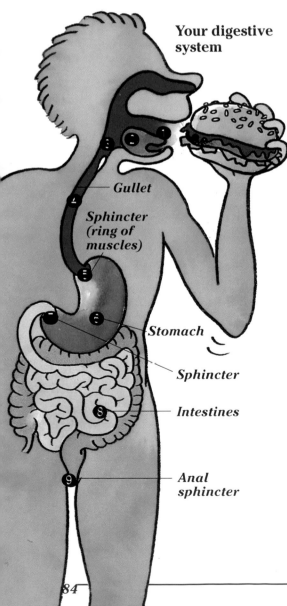

Your digestive system

— Gullet

Sphincter (ring of muscles)

— *Stomach*

— *Sphincter*

— *Intestines*

— *Anal sphincter*

❶ Food is put into mouth.

❷ Food is chewed and shaped into a ball, or bolus, by tongue.

❸ Throat muscles open and you swallow. Bolus enters gullet.

❹ Bolus is pushed along by muscles in the gullet by a process called peristalsis.

Peristalsis

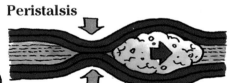

Muscles contract.   Food moves.

❺ Sphincter muscle at entrance to stomach relaxes (opens) so food enters stomach.

❻ Stomach muscles contract and relax. This churns food, mixing it with chemicals that break it down.

❼ Sphincter muscle at stomach's exit relaxes (opens) and food enters intestines.

❽ Food is pushed along by peristalsis. Digested food is absorbed into blood stream.

❾ Unwanted food is pushed out through the anal sphincter when you go to the toilet.

Which of these stages are usually controlled by involuntary muscles? (Answer on page 98.)

BREATHING

When you breathe, muscles force air in and out of your lungs. This is usually involuntary, but you can control your breathing if you want to.

Breathing in   *Air*

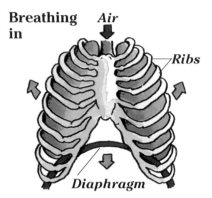

Ribs

Diaphragm

1. An arched sheet of muscle under the lungs, called the diaphragm, contracts and flattens.

2. Muscles between the ribs contract, pulling the ribs up and out.

3. The lungs expand so air rushes into them to fill the space.

Air

Breathing out

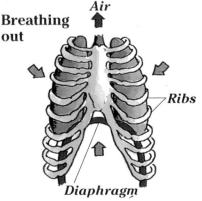

Ribs

Diaphragm

1. The diaphragm relaxes, so becomes arched again.

2. Rib muscles relax so the ribs move down and in.

3. Air is squeezed out of the lungs.

MUSCLE VISION

Without the work of involuntary muscles you wouldn't be able to see the things around you. When you look at an object, an image of it is created on the retina at the back of your eye. Light rays from the object travel into your eye through a hole called the pupil. A ring of muscle called the iris opens and closes the pupil. Then the rays pass through a lens which focuses them so they form a clear image on the retina. Tiny muscles, called ciliary muscles, change the shape of the lens to focus the rays (see page 38).

This diagram shows a side view of the inside of an eye.

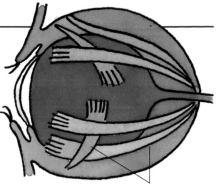

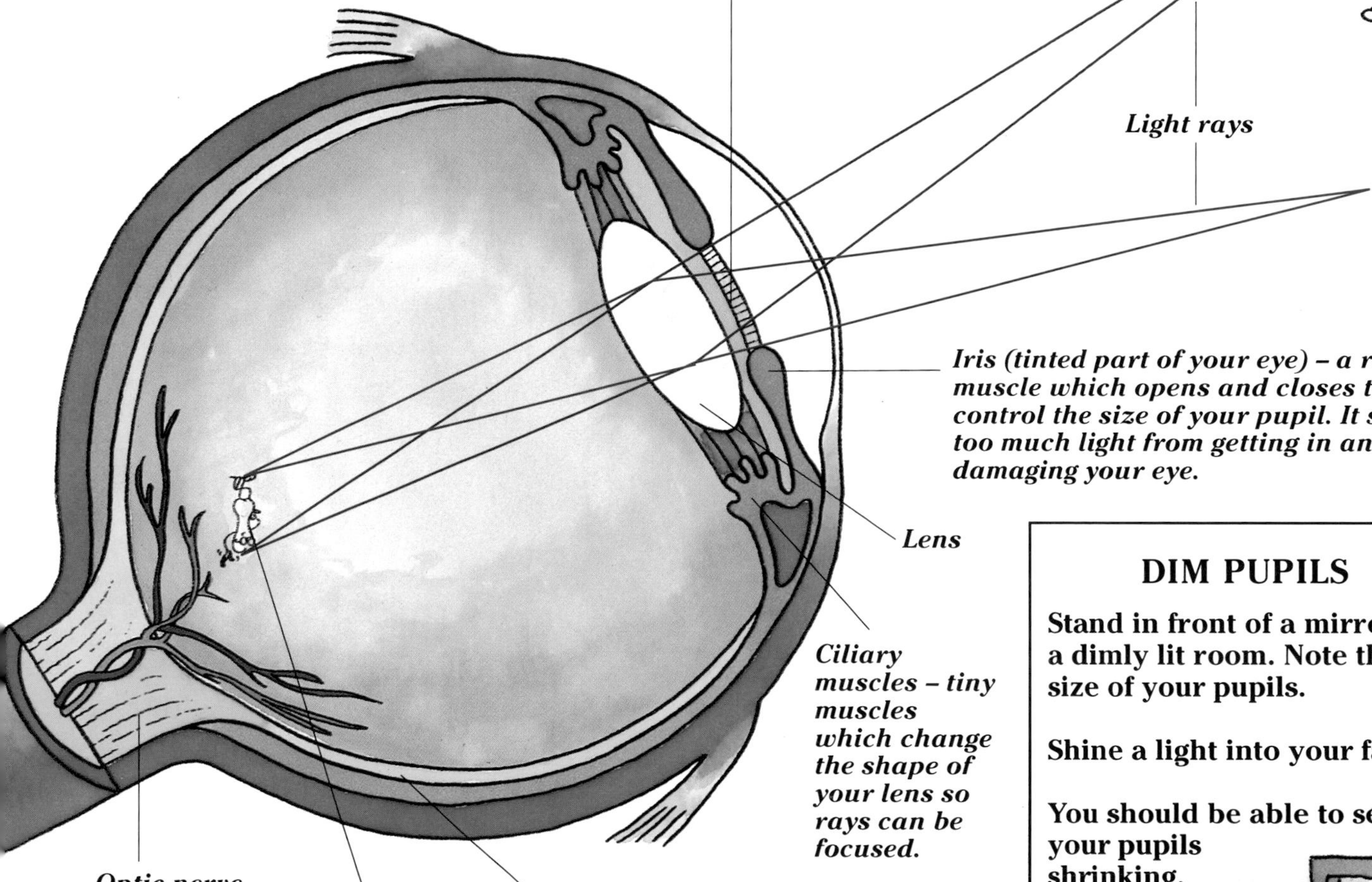

Eye socket muscles attach eyeball to skull.

You can't feel it, but these involuntary muscles make your eyeballs flicker constantly in your eye socket. You can take over control when you want.

Pupil – hole through which light enters your eye. It looks like a black dot. It gets bigger in the dark and shrinks in the light.

Light rays

Iris (tinted part of your eye) – a ring of muscle which opens and closes to control the size of your pupil. It stops too much light from getting in and damaging your eye.

Lens

Ciliary muscles – tiny muscles which change the shape of your lens so rays can be focused.

Optic nerve sends impulses to your brain.

Image on retina

Retina – layer of tissue at the back of your eyeball which contains millions of nerve cells.

DIM PUPILS

Stand in front of a mirror in a dimly lit room. Note the size of your pupils.

Shine a light into your face.

You should be able to see your pupils shrinking.

MUSCLE TALK

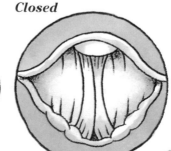

You can speak. Speech is a series of different sounds with meanings. Human speech is very complicated and sets us apart from all other animals. In order to speak, you have to be able to control a system of different muscles.

WINDBAGS

In order to make any sound at all, you first have to have a stream of air. When you speak, you use the air you breathe out of your lungs. (You can use the air you breathe in, but it will be muffled and croaky. Try it and listen.)

Actors have special training to help them control their breathing.

Breathing is controlled by the plate of muscle under your lungs called the diaphragm.

Breathe in for 2½ seconds. **Breathe out for 5-10 seconds.**

Air in *Air out*

Diaphragm

Usually you spend the same amount of time breathing in as you do breathing out - about 2½ seconds. But when you speak, you use your diaphragm to make the breathing out time last about 5-10 seconds. This means you can get more speaking time out of each breath.

VIBRATING VOICE BOX

When air leaves your lungs, it enters your voice box, or larynx (see diagram on page 87). Within your voice box are two muscly bands called vocal cords. These open and close rapidly, making the air in your voice box vibrate. These vibrations can be heard as sounds. The faster your vocal cords open and close, the higher the sound you will make. Women's vocal cords

open and close about 220 times a second, and men's about 120 times a second. This is why men have deeper voices.

View of vocal cords from above

Open *Closed*

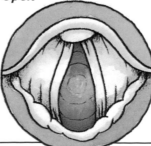

Sopranos (female singers) sing very high notes. Their vocal cords open and close up to 1000 times a second.

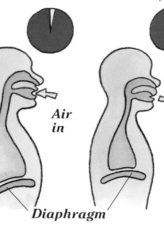

OOH, AHH, MUSCULAR

Air passes from your voice box into your vocal tract (see diagram on right). From here it leaves your body through your mouth or nose. On its way out, the air is affected by several muscles which work together to produce a wide range of speech sounds.

Say these words and see how the position of your lips changes:

cut

cat

curt

cool

Your tongue is an incredibly flexible muscle, which can bend and curl into many different positions. It is vital for making nearly all speech sounds.

To see how important your tongue is, try to read this sentence out loud while holding down your tongue with your finger.

urmm yuurm thith...wuuth...

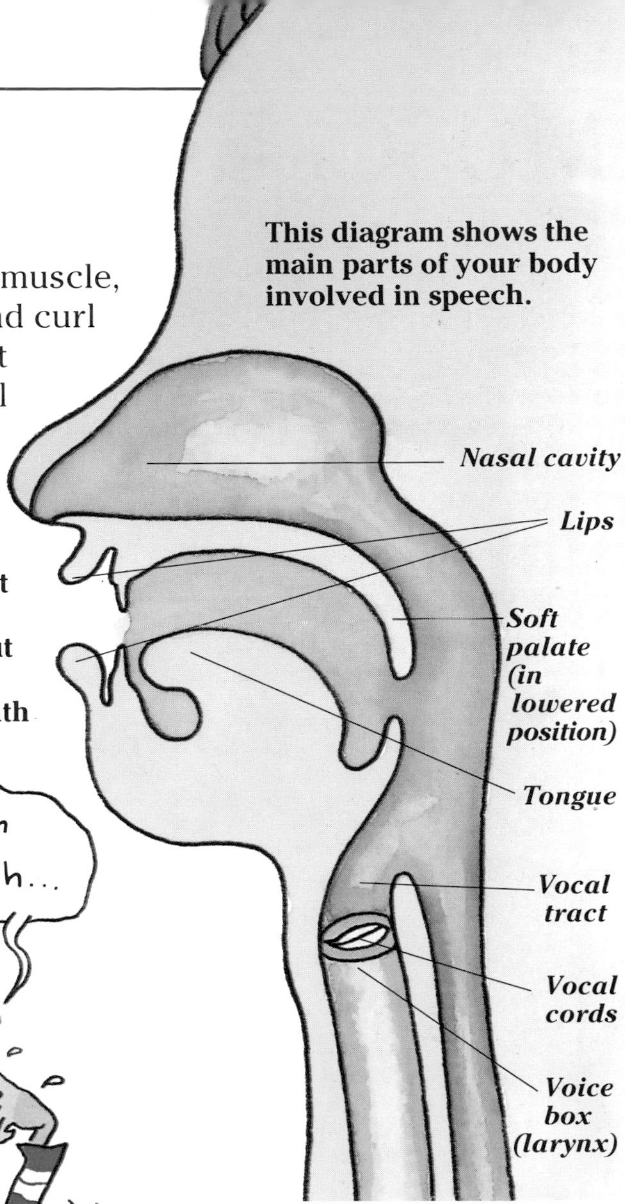

This diagram shows the main parts of your body involved in speech.

Nasal cavity

Lips

Soft palate (in lowered position)

Tongue

Vocal tract

Vocal cords

Voice box (larynx)

Your lips are a ring of muscly tissue. You form them into different shapes for different sounds. They are completely closed for b, p and m sounds, and open different amounts for all the vowel sounds.

Your soft palate is a band of muscle at the back of your mouth. You can move it up and down to control how the air leaves your body. When it is up, air can escape only through your mouth. It is up for most sounds. When it is lowered, air can escape through your nose too. It is like this for nasal sounds, such as m, n and the 'oi' sound in 'oink'.

TONGUE TWISTERS

Tongue twisters are phrases which are difficult to say clearly and quickly.

Try to say these tongue twisters over and over again, as fast as you can.

"Red lorry, yellow lorry"

"Peggy Babcock, Peggy Babcock"

GROWING

As you get older you grow taller because your bones get longer. The bones that grow the most are your leg bones - the femur, tibia and fibula. Most people have stopped growing by their early twenties. Some aren't happy with their final height and want to be taller or shorter. But once you have stopped growing, there is nothing you can do to change it.

BOUNCING BABES

Before it is born, a baby develops a skeleton. This skeleton is not made of bone, but of softer, gristly stuff called cartilage. (This is the same as the cartilage which cushions your joints.) After the baby is born, the cartilage slowly begins to turn into bone. This process is called ossification. By about the age of 12 nearly all your skeleton will be ossified.

A baby's cartilage skeleton makes it more flexible than older children and adults. They are less likely to break their bones.

GROWTH PLATES

Your bones cannot grow longer simply by making new bone cells. First they have to make new cartilage which is then turned into bone. Once your skeleton has ossified, some bones keep two small pads of cartilage. These are special growth plates (see right), which allow you to continue growing. In each plate the inner edge of the cartilage is gradually ossified. At the same time new cartilage grows at the outer edge. When growing stops completely, these plates of cartilage finally become bone themselves.

Child's bone - growth takes place throughout

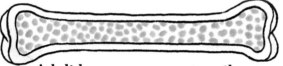

Growth plates

Adolescent's bone - growth only at growth plates

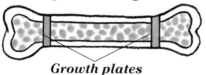

Adult bone - no more growth

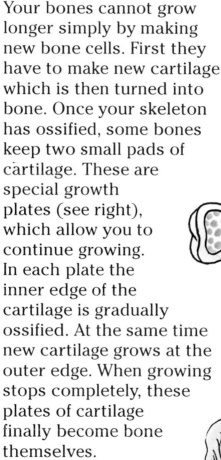

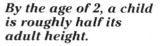

A baby grows most in the first year of life.

By the age of 2, a child is roughly half its adult height.

Most children grow an average of 5-7.5cm (2-3in) a year.

GROWTH CONTROL

When and how much you grow is controlled by hormones. Hormones are chemicals released by organs in your body, called glands. They carry instructions to your cells, telling them what to do. Growth hormones are produced by a gland in your brain called the pituitary gland.

At puberty, when a child becomes sexually mature, sex hormones are also released. These lead to a growth spurt. They also lead to changes in the skeleton which are different for men and women.

Pituitary gland

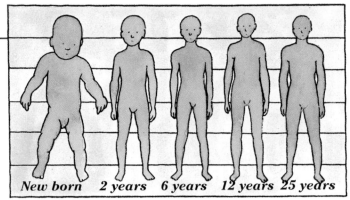

New born   2 years   6 years   12 years   25 years

Different parts of your body grow at different ages. This chart shows how your body proportions change as you grow up.

Female sex hormones, called estrogens, make a woman's pelvis grow wider to make it easier for her to have children. The male sex hormone, testosterone, causes men's bones to grow larger and heavier than women's.

CRUMBLY BONES

Sex hormones also make sure your bones repair themselves and absorb enough minerals to keep them strong. Old people produce fewer sex hormones, which means their bones are weaker. This can lead to a disease called osteoporosis, which makes bones crumbly, fragile and more breakable.

Most girls start puberty between the ages of 10 to 14. (Boys between 12 to 16.) It lasts for about 3-4 years, until maturity is reached.

Old people may stoop as their bones get weaker.

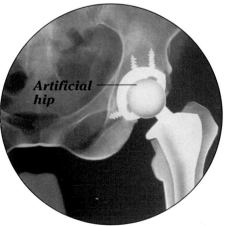

Artificial hip

People can have an operation to replace a weak hip with an artificial one.

89

BROKEN BONES

Although your bones are strong, they can break if you put too much weight on them or twist a joint the wrong way. Broken bones are called fractures. Some fractures are more serious than others.

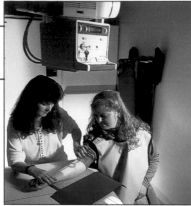

Girl having an arm x-ray

X-RAYS

If doctors suspect you have a fracture, they take an x-ray. An x-ray machine sends a beam of rays through your body. The rays pass through your skin, fat and muscle to a special photographic plate, which turns dark. But the rays cannot pass through your bones, so your bones show up on the plate as light areas. From this, doctors can tell if you have a fracture and how serious it is.

The x-rays on the right show different types of fractures. They are tinted to make them clearer.

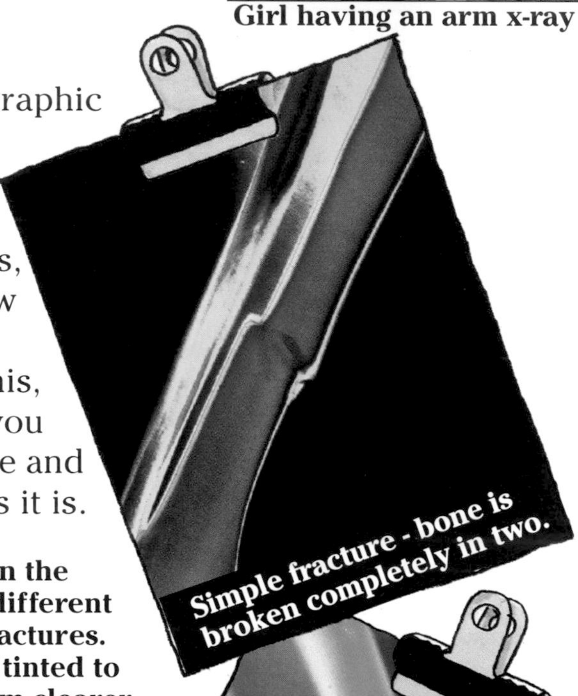

Simple fracture - bone is broken completely in two.

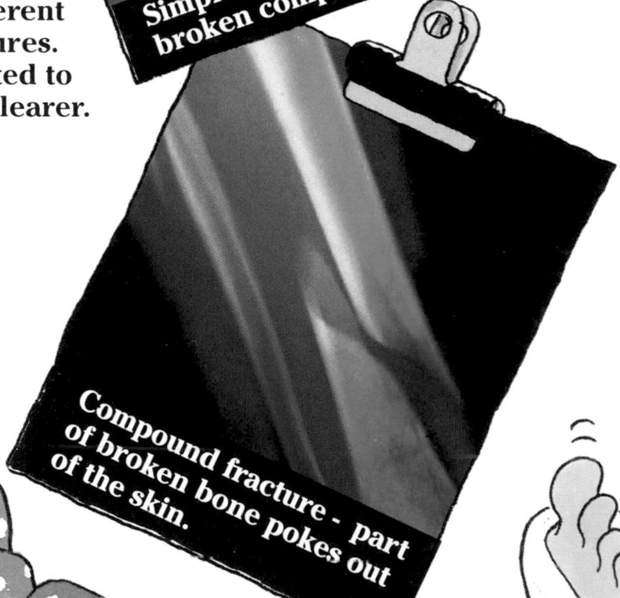

Compound fracture - part of broken bone pokes out of the skin.

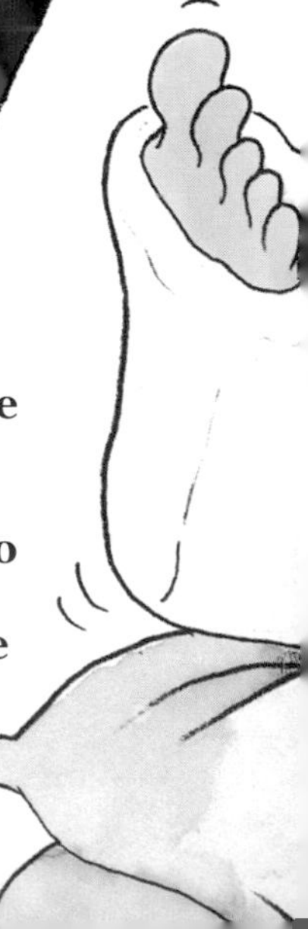

To help a bone repair itself, doctors use plaster casts to keep the pieces of bone in place.

90

For a link to a website where you can find out more about what happens if you break a bone, go to **www.usborne-quicklinks.com**

MENDING BONES

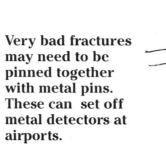

With a bit of help from doctors, your bones will mend themselves, but it can take several months. This is a long time compared with how long it takes your skin to heal. It only takes a few days for new skin to grow over a cut.

Very bad fractures may need to be pinned together with metal pins. These can set off metal detectors at airports.

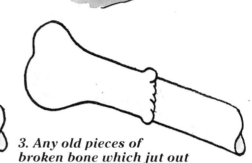

A bone mends itself in three stages:

1. When a bone fractures, blood vessels are broken too. Blood pours from the torn vessels and forms a clot, around the broken ends of bone.

2. New bone cells move into the blood clot. As more bone cells are made, the blood clot is gradually replaced by bone.

3. Any old pieces of broken bone which jut out are absorbed into the blood stream, so the repair is smoothed down.

BONE MARROW DISEASE

Leukemia is a disease of the bone marrow. It upsets the production of healthy white blood cells. These are the very cells that fight disease. There are several different types of leukemia. In one type, the marrow produces too many white blood cells, so the red ones are swamped. In another, white blood cells are let into the blood too early, before they are ready to carry out their disease-killing tasks. In a third type, the body is overrun with useless old white blood cells which should have died.

No one knows exactly what causes leukemia, but most cases can be treated with drugs or radiotherapy. In radiotherapy the problem white blood cells are zapped with high doses of x-rays in order to kill them. Many patients make a full recovery.

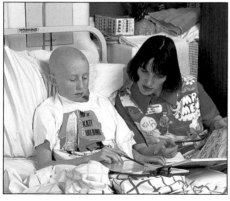

A leukemia patient. The drugs that treat leukemia may make your hair fall out, but they can cure the disease completely.

MUSCLE TROUBLE

Muscles often cause aches and pains, but this is usually nothing serious. You may have "pulled" a muscle by making it stretch too violently or unexpectedly, or strained muscles which don't usually have to work hard. Rest and a relaxing hot bath are often the best things for this sort of muscle pain. There are, however, more serious things that can go wrong with your muscles.

Nerve cell in brain

DYSTROPHIES

Inside all your cells are chains of chemicals called genes. Genes carry instructions which tell your cells what to do. Particular genes in your muscle cells are essential for your muscles to contract. In some people these genes are faulty and they suffer from diseases called muscular dystrophies. Their muscles gradually stop working and waste away. Most sufferers end up in a wheelchair and many die at a young age.

At the moment there is no cure for dystrophies. Scientists are hoping that one day they may be able to correct bad genes or replace them with healthy ones. But this is still a long way in the future.

Scientists need very strong microscopes to study genes.

PARALYSIS

Muscles contract when they receive messages from the brain. So, if someone damages their brain, or the nerves which connect it to the rest of their body, this can also affect their muscles. They could suffer paralysis and be unable to move their muscles.

Part of the brain which controls movement

Brain damage is not necessarily permanent. Sometimes patients regain feeling and movement in parts of their bodies. Doctors don't really know fully how the brain works, so no one understands quite how or why this happens.

It's very important to wear a helmet when you are cycling so your head is protected if you fall off.

MYELIN AND MULTIPLE SCLEROSIS

Messages, or electrical pulses, from your brain travel to your muscles via nerve cells (see page 6). Nerve cells are covered with a layer of fat called a myelin sheath.

In a disease called multiple sclerosis (MS), these fatty layers are destroyed. Without myelin sheaths, nerve cells can't transmit messages properly.

This means the messages that reach the muscles are not as clear and strong as they should be. So sufferers gradually lose the use of their muscles. No one has yet discovered the cause of MS. Doctors have come up with theories linking it to climate, diet, genes and viruses, but none of them has been proved. Although some of the symptoms of MS can be treated with drugs, there is no actual cure.

This diagram shows a nerve cell damaged by multiple sclerosis.

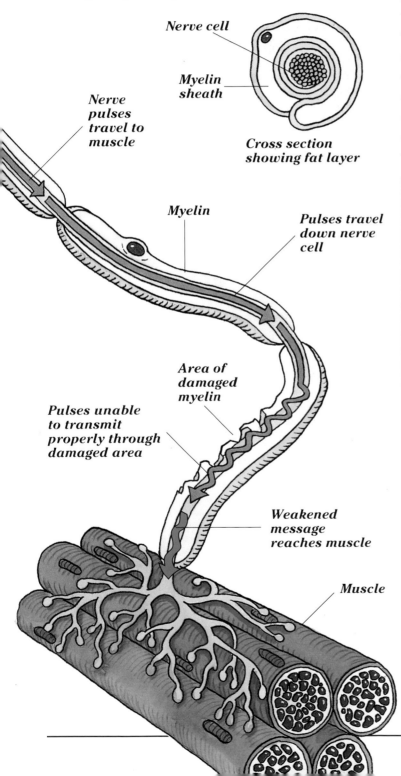

Nerve cell

Myelin sheath

Nerve pulses travel to muscle

Cross section showing fat layer

Myelin

Pulses travel down nerve cell

Area of damaged myelin

Pulses unable to transmit properly through damaged area

Weakened message reaches muscle

Muscle

PHYSIOTHERAPY

If muscles are not used, they very quickly become flabby and weak. So, if, for example, you have been ill in bed for a long time, you will need to do special exercises to build up your strength gradually. These are called physiotherapy. There are many different types of physiotherapy, such as massage, heat treatment and swimming.

PREHISTORIC BONES

Evolution is the theory that living things have changed, or evolved, over millions of years. Many scientists believe that the animals that live on the earth today are descended from simpler forms that lived in earlier ages. Much of the evidence for evolution lies in the fossils of old bones, some many millions of years old.

BURYING THE EVIDENCE

Fossils are the remains of animals or plants which have been turned into rock. Bones are fossilized when they are covered with mud or sand which, over millions of years, is compressed to form rock. Tiny rock particles enter the bones which become fossils.

Ramapithecus – *ape living 12 million years ago. First apes to leave trees.*

Australopithecus afarensis – *ape living 4–3 million years ago. First apes to use sticks and stones.*

Homo habilis (handy human) – *living 2–1.5 million years ago. First to make tools.*

HUMAN EVOLUTION

Scientists have found very few fossils of the ancestors of human beings. But they have built up a picture of how they think humans evolved, based on skulls and teeth which have been unearthed. As new evidence is discovered, new theories of human evolution are put forward.

Homo erectus (upright human) – *living 1.6 million–400,000 years ago, probably spread from Africa to Europe. First to use fire.*

Homo sapiens sapiens - *first modern human, evolved 100,000 years ago.*

This time line shows how humans may have evolved.

These skulls show how the shape of our heads has changed over millions of years.

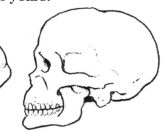

Skull of Australopithecus

Skull of Homo erectus

Skull of modern human

AMAZING FACTS

Robert Wadlow with his brother

The largest recorded human bone belonged to a German giant called Constantine. His femur was 76cm (29.9in) long.

Giantism and dwarfism are medical conditions. They are caused when the pituitary gland, which makes growth hormones, malfunctions.

The strongest muscle in your body is the jaw muscle which you use for biting.

The tallest recorded man in the world ever was Robert Wadlow. He was born in 1918 and died aged 22, measuring 272cm (8ft 11.1in). He hadn't reached his full height when he died and doctors think he could have grown to 274cm (9ft).

Gul Mohammed

The shortest recorded fully-grown man in the world is the dwarf Gul Mohammed of New Delhi, India. He was measured in 1990 and was only 57cm (22½in) tall.

The muscle with the longest name is the *levator labii superioris alaeque nasi*. This is Latin. The muscle is in your face and it enables you to curl your upper lip and twitch your nose.

Elvis Presley – one of the world's most famous lip curlers

The largest muscle in your body is in your bottom. But, in pregnant women, the muscular womb (where a new baby grows) can grow bigger than the bottom.

Your smallest muscle is less than 0.127cm (0.05in) long. It is attached to a tiny bone in your ear.

The most active muscles in your body are your eye muscles. They contract about 100,000 times a day. Much of this movement takes place when you are dreaming and your eyes flicker under your closed eyelids.

DID YOU KNOW?

All mammals, except sea-cows and sloths, have seven bones in their necks. This includes giraffes whose necks can be 2m (6.5m) long and mice who seem to have no neck at all.

For a link to a website where you can play a game to see how your skeleton fits together, go to **www.usborne-quicklinks.com**

INDEX

Achilles heel, 71
Achilles tendon, 71
aerobic respiration, 80, 81
anaerobic respiration, 80, 81
alcohol, 24
amplitude, 43, 64
Ancient Greeks, 30
animal brains, 26-27
animal senses, 56-57
archeologists, 77
Aristotle, 30
arthritis, 73
ATP, 80, 81
auditory nerve, 42
axon, 6, 7, 9 36, 37

babies, 8-9, 67
babies' senses, 58, 59
balance, 34,.37, 53
biceps, 70, 74
binary code, 62
Binet, Alfred, 10
binocular vision, 12, 39, 41
blindness, 44, 54-55, 64
blind spot, 38, 55
blood, 7, 19, 70, 76, 78-79, 80,
    84, 91
blood cell, 19
bone,
    biggest in body, 69
    broken, 90-91
    cells, 67, 76, 88, 91
    compact, 76
    joints, 72-73, 90
    marrow, 76, 91
    smallest in body, 68
    spongy, 76
    tissue, 67, 76
Braille, 48
brain, 36, 37, 38, 39, 40, 41, 42,
    47, 48, 50, 51, 52, 53, 54, 55,
    58, 59, 62, 67, 68, 74, 75, 78,
    84, 85, 89, 92
breathing, 80, 84

calcium, 76, 77
car sickness, 53
cartilage, 72, 73, 88
cataracts, 44
cell body, 6, 7, 36
cells, 67, 74, 75, 76, 77, 80, 85,
    88, 91, 92, 93

cerebellum, 4
cerebral hemispheres, 4, 5, 10, 32
cerebrum, 4, 26
cochlea, 42
color, 37, 39, 40, 56
colorblindness, 39, 56, 64
computers, 62-63
cones, 12, 13, 38, 39, 64
consciousness, 20-21, 32
constancy mechanisms, 40, 64
contracting, 74, 75, 77, 82, 83
coordination, 67
corpus callosum, 4, 5
cortex, 4, 5, 36, 64
craniologist, 3, 32

deafness, 44
decibels, 43, 64
delusions, 22
dendrite, 6, 7, 9, 36, 37
depression, 22, 24
diaphragm, 84, 86
digestive system, 84
dizziness, 53
dopamine, 23
dreams, 21
drugs, 23, 24, 51
dwarfism, 95

ear, 35, 37, 42-43, 44, 57
    inner ear, 42, 53
    middle ear, 42
    outer ear, 42
ear drum, 42
echo location, 57
EEG (electroencephalograph),
    21
electrical pulses/signals, 6, 7, 12,
    15, 21, 29, 36, 37, 38, 42, 44,
    48, 53, 60, 62, 63
endorphins, 51, 64
ESP (extra sensory perception),
    25
evolution, 69, 94
exercise, 73, 83
exoskeleton, 69
eye, 12-13, 37, 38-39, 44, 57, 85
    95

feeling (see touch)
feelings (emotions), 6
flavors, 46, 47

focusing, 38
fossils, 94
fractures, 90, 91
Freudian slip, 21
Freud, Sigmund, 20, 21

Galen, 30
genes, 11, 23, 32
giantism, 95
gray matter, 7
growing, 88-89
growth, of brain, 9

hair cells, 42, 53
hallucinations, 22, 24
hallucinogens, 24
harmonics, 43, 64
hearing, 4, 6, 8, 12, 34, 42-43, 59
hearing aids, 44, 45
heart, 68, 78-79, 80, 81
homeostasis, 18-19, 32
Homer, 30
hormones, 18, 19, 89
Hubble telescope, 62
hypnosis, 25
hypothalamus, 4, 18, 19, 32

illusions, 40, 60-61
incus, 42
intelligence, 10-11
    in animals, 26, 27
    in computers, 28, 29
invertebrates, 69
involuntary actions, 75, 84, 85
involuntary muscles, 84-85
IQ tests, 10, 11
iris, 85

joints, 37, 52, 72-73, 90

kidneys, 19

lactic acid, 81
language, 5, 8
larynx, 86, 87
learning, 8
    in animals, 26
left side of brain, 5
lens, 38, 85
leukemia, 91
ligaments, 72, 73
light, 37, 38, 39

lip reading, 45
localization, 31
lungs, 68, 78, 79, 80, 84, 86

malleus, 42
marrow, 76, 91
memory, 14-15, 16-17, 26
        and dreams, 21
mental illness, 22-23
migration, 57
morphine, 51
movement, 4, 6, 67, 70, 73, 74-75,
        81, 82
multiple sclerosis, 93
muscle,
        biggest in body , 95
        cells, 67, 77, 80
        digestive, 84
        fatigue, 70, 81
        hand, 66
        heart, 77, 78-79
        involuntary, 50, 84-85
        most active, 95
        skeletal, 70, 74, 77,
        smallest in body, 95
        stomach, 70, 84
        striped, 77
        strongest, 95
        tissue, 67, 77
        voluntary, 52, 70
muscular dystrophies, 92

nerves, 74, 75, 76, 85, 92, 93
nervous system, 7, 29, 30, 32
neurologist, 3, 9, 32
neurons, 6, 7, 9, 32, 36-37, 50
        and memory, 15, 23
nose, 37, 47, 57

olfactory bulb, 36
optical illusions, 13
optic nerve, 12, 38
ossification, 88
osteoarthritis, 73
osteologists, 67
osteopaths, 67
osteoporosis, 89
oval window, 42
oxygen, 7, 19, 23, 70, 79, 80, 81

pain, 37, 48, 50-51
        referred, 50, 64

painkillers, 24, 51
paralysis, 92
Parkinson's disease, 23
phobias, 22
phrenology, 31, 32
physiotherapy, 67, 93
pitch, 43, 60, 64
planning, 4
pons, 4, 5
pressure, 37, 48, 49
psi, 25, 32
psychiatrist, 3, 32
psychokinesis, 25
psychologist, 3, 11, 32
pulse, 79
pupil, 38, 39

receptors, 36, 38, 42, 46, 47, 48,
        50, 53, 55, 60, 62, 64
        cold, 48, 49
        free nerve endings, 48
        hair cell, 42, 53
        hair root, 48
        heat, 48, 49, 57
        Pacinian, 48
        pressure, 48
        touch, 48, 49
referred pain, 50, 64
reflex actions, 50, 75
REM, 21, 32
remembering, hints and cues, 16-
        17
respiration, 80-81
retina, 12, 38, 55, 61
retinal image, 12, 13, 32
        38, 40, 41, 52
rheumatoid arthritis, 73
right side of brain, 5
rods, 12, 13, 38, 39, 64

saliva, 37, 46
scan, brain, 23, 31
scanners, 62
schizophrenia, 22, 32
sedatives, 24
seeing, 4, 6, 8, 12-13, 34, 35, 38-39,
        40-41, 52, 54-55, 58
        in 3-D, 12, 38, 39, 41, 61
semicircular canals, 53
sense organs, 34, 36, 48
sight (see seing)
sign language, 45

sixth sense, 35
skeletal muscles, 70, 74, 77
skeleton, 68-69, 70, 77, 88
skin, 37, 48
skull, 68, 85, 94
sleep, 21
smell, 5, 6, 8, 12, 34, 35, 36, 37,
        46-47, 56, 59
sound waves, 42, 43, 44, 61, 63
sounds, 36, 37, 43
speech, 4, 5, 8, 86-87
sphincters, 84
spinal cord, 4, 7, 29, 50, 51, 74, 75
spine, 68, 69
sports, 82-83
stapes, 42
stimulants, 24
strength, 82-83
stroke, 23
synapses, 7, 32
synovial fluid, 72, 73
synovial membrane, 72

taste, 6, 12, 34, 35, 46-47, 59
temperature, 37, 48, 49
tendons, 52, 71, 73
thalamus, 4, 36, 64
therapy, 23
thinking and thoughts, 5, 6
        in children, 8, 9
        in pictures, 5, 9
        in the unconscious, 20
tissue, 67, 77
tongue, 37, 46, 47, 87
touch, 4, 12, 34, 35, 36, 37, 48-49
touchscreens, 63
training, 83
trepanning, 31
triceps, 70, 74
twins, identical, 11

unconscious, 20, 21

vertebrates, 69
vestibular system, 53, 64
virtual reality, 63
vocal cords, 86, 87
voice box, 86, 87
voluntary actions, 75
voluntary muscles, 70

x-rays, 90, 91

QUIZ ANSWERS

Page 5
Right or Left?
1. d (Right)
2. 5 (Left)
3. c (Right)
4. Mary (Right)

Pages 10 and 11
Taking the test
1. 22
2. 3
3. The middle shape
4. 5
5. Short
6. Bird
7. Yes

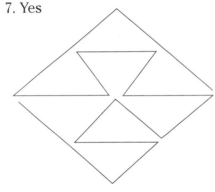

8. The second piece from the left.
9. 6

Page 61
Perspective illusions
1. They are both the same length.
2. They are all the same height.
Where's the square?
3. All of them.

Page 68
Necktie
The same number- 7!

Page 82
Quiz
1. Maximum strength
2. Specific strength and endurance
3. Explosive strength

Page 84
Digestive muscles
4, 5, 6, 7 and 8

ACKNOWLEDGEMENTS

Page 2: Petit Format/J.D. Bauple/ Science Photo Library (SPL). Page 10: Oxford and County Newspapers. Page 13: Man and Girl series of ambiguous figures - Fisher, G.H. (1967) From "Perception of ambigous stimulus materials," Perception and Psychophysics, 2:421-22. Reprinted by permission of Psychonomic Society, Inc. Page 20: National Library of Medicine/ SPL. Page 23: Scan comparing two brains - U.S. National Institute of Health/SPL; Scan showing blocked artery - CNRI/ SPL.
Page 28: Artoo Deetoo - ™ & © Lucasfilm Ltd. (LFL) 1980. All Rights Reserved. Courtesy of Lucasfilm Ltd. BFI Stills, Posters and Designs; Robotic welding machine - George Haling/SPL.

Page 29: Rex Features Ltd - Assignments. Page 31: BSIP DPA/SPL. Page 44: Isabelle Lilly, Photographer (with thanks to the RNIB). Page 51: Corbis-Bettmann. Page 56: David Scharf/Science Photo Library. Page 58: Sue Atkinson, Photographer (with thanks to Charles York-Miller). Page 61: Mary Evans Picture Library (After Boring, 1930). Page 62: spiral galaxy - Space Telescope Science Institute/NASA/Science Photo Library; computer processor - Alfred Pasieka/ Science Photo Library. brain cross section - Scott Camazine/ Science Photo Library. Page 67: Bone tissue, Michael Abbey/ Science Photo Library (SPL). Page 71: © Usborne Publishing Ltd. Photography by Howard Allman. Page 73: CNRI/SPL. Page 74: Don Fawcett/SPL. Page 75: Anatomical studies, The Royal Collection©, Her Majesty Queen Elizabeth II. Page 77: Photo: Museum of London Archaeology Service. Page 78: John Radcliffe Hospital/SPL. Page 82: Gray Mortimer/ Allsport. Page 89: Chris Bjornberg/SPL. Page 90: Radiographer - Will and Deni Mcintyre/SPL; Simple fracture - SPL; Compound fracture - SPL. Page 91: Simon Fraser/Royal Victoria Infirmary/Newcastle/ SPL. Cover: John Durham/SPL